INTERMEDIATE 2

PHYSICS

Hugh McGill

Hodder Gibson
A MEMBER OF THE HODDER HEADLINE GROUP

Every effort has been made to trace all copyright holders, but if any have been inadvertently overlooked the Publishers will be pleased to make the necessary arrangements at the first opportunity.

Although every effort has been made to ensure that website addresses are correct at time of going to press, Hodder Gibson cannot be held responsible for the content of any website mentioned in this book. It is sometimes possible to find a relocated web page by typing in the address of the home page for a website in the URL window of your browser.

Hachette's policy is to use papers that are natural, renewable and recyclable products and made from wood grown in sustainable forests. The logging and manufacturing processes are expected to conform to the environmental regulations of the country of origin.

Orders: please contact Bookpoint Ltd, 130 Milton Park, Abingdon, Oxon OX14 4SB. Telephone: (44) 01235 827720. Fax: (44) 01235 400454. Lines are open 9.00 – 5.00, Monday to Saturday, with a 24-hour message answering service. Visit our website at www.hoddereducation.co.uk. Hodder Gibson can be contacted direct on: Tel: 0141 848 1609; Fax: 0141 889 6315; email: hoddergibson@hodder.co.uk

First published in 2008 by
Hodder Gibson, an imprint of Hodder Education, part of Hachette Livre UK,
2a Christie Street
Paisley PA1 1NB

Impression number 5 4 3 2 1
Year 2012 2011 2010 2009 2008

Cover photo David Parker/Science Photo Library
Cartoons by Moira Munro/Artworks by Phoenix Photosetting
Typeset in 9.5 on 12.5pt Frutiger Light by Phoenix Photosetting, Chatham, Kent
Printed in Great Britain by Martins the Printers, Berwick upon Tweed

A catalogue record for this title is available from the British Library.

ISBN-13: 978 0340 946 749

CONTENTS

WELCOME

Welcome to this revision book!

And welcome to Intermediate 2 (Int2) Physics! This book is designed to help you.

Int2 Physics is enjoyable, interesting, and relevant to life in the twenty-first century. It is also very useful if you want to progress to Higher Physics, apply to college, or want to get a job.

The examination is fair. Prepare properly for it and you will do well.

Good luck!

How to use this book

This book covers everything you need for Int2 Physics. It also includes hints and tips about how to answer exam questions and about what SQA expects you to do.

The book has six Chapters.

The first Chapter covers skills which you need for the course – general things like units, prefixes, and scientific notation. It also includes advice on how to prepare for the exam. You should refer to this Chapter often when you are preparing for your prelim and for the final examination.

The next four Chapters cover the knowledge and understanding of the four Int2 units: Mechanics and Heat, Electricity and Electronics, Waves and Optics, and Radioactivity.

Each Chapter is written in topics. At the start of each topic there is a summary of what you need to be able to do. You can use these summaries to help you prepare a study checklist for your prelim, or for the final examination. Study the topics one at a time – in any order you like.

Some topics include examples that show you how to answer exam questions. All of the topics in Chapters 2 to 5 include questions for you to try. Just like the final examination, some of the questions are straightforward and some are more difficult.

Detailed answers to all of the questions are included in Chapter 6. For maximum benefit, try to answer the questions without looking at the answers. However, if you do not know how to tackle a question, the answers are there to help you. Use the answers wisely and you will learn a lot!

Study all of the topics and try all of the questions before you sit the final examination.

Chapter 1

COURSE SKILLS

1.1 Units, prefixes, and scientific notation

What You Should Know

For Intermediate 2 Physics you need to be able to:

- use SI units of all physical quantities included in the course
- give answers to calculations to an appropriate number of figures
- use scientific notation
- understand and use the prefixes μ, m, k, M, G.

Units

You have to be able to use the SI units of all physical quantities included in the Int2 Physics course. A full list of quantities is included in the table at the end of this topic.

Get into the habit of using the correct SI units every time you tackle a numerical problem. Do this, and by the time you sit the examination you will get units correct without having to think.

In Int2 Physics there is usually ½ mark for the correct unit in the **final answer** to a numerical problem. It may not seem much, but half marks can accumulate into enough to make a difference between passing and failing.

You do not have to write units in the middle of a calculation. Your statement of the final answer **must** include the correct unit.

Significant figures

The number of significant figures in your **final answer** to a numerical question (or part of a question) should be the same as the **minimum** number of significant figures in the data you use to work out the answer.

Do not round intermediate values to the number of significant figures for the question. Keep one additional figure in intermediate values and **round when you write your final answer**. This method is used in the examples and the solutions to the exercises in this book.

Be careful about significant figures when you use data from the data sheet at the front of the exam paper. For example, gravitational field strengths may be quoted to only one figure.

Having too many figures in final answers can cost you marks – do not let this be the reason you do not pass your Int2 Physics!

Scientific notation

Some data in this course are very large or very small numbers. You have to understand and use the notation for these numbers. For example,

Specific latent heat of fusion of water = $3{\cdot}34 \times 10^5$ J/kg.

Make sure you practise this skill often!

Prefixes

The following table includes all the prefixes that you need to understand for Int2 Physics. Learn them and get into the habit of using them.

Prefix	Short for	Means	Prefix	Short for	Means
m	milli	$\times 10^{-3}$	k	kilo	$\times 10^3$
μ	mu	$\times 10^{-6}$	M	mega	$\times 10^6$
			G	giga	$\times 10^9$

The kilogram is the SI unit of mass – **do not** change kilograms to grams.

Physical quantities in Intermediate 2 Physics

The following table shows all of the quantities, their symbols, units, and abbreviations, that you will meet in Int2 Physics.

Physical Quantity	Symbol	Unit	Abbreviation
distance, displacement	s or d	metre	m
height	h	metre	m
time	t	second	s
speed, instantaneous speed, velocity, final velocity	v	metre per second	m/s
initial velocity	u	metre per second	m/s
change of velocity	Δv	metre per second	m/s
average velocity, average speed	$\bar{v}$	metre per second	m/s
acceleration	a	metre per second per second	m/s^2
acceleration due to gravity	g	metre per second per second	m/s^2
gravitational field strength	g	newton per kilogram	N/kg
mass	m	kilogram	kg
force, resultant force	F	newton	N
momentum	p	kilogram metre per second	kg m/s
weight	W	newton	N
energy	E	joule	J
work done	E_W	joule	J

Physical Quantity	Symbol	Unit	Abbreviation
potential energy	E_p	joule	J
kinetic energy	E_k	joule	J
power	P	watt	W
percentage efficiency	-	-	-
output energy	E_o	joule	J
input energy	E_i	joule	J
output power	P_o	watt	W
input power	P_i	watt	W
heat energy	E_h	joule	J
temperature	T	degree celsius	°C
specific heat capacity	c	joule per kilogram per degree celsius	J/kg °C
specific latent heat	l	joule per kilogram	J/kg
electric charge	Q	coulomb	C
electric current	I	ampere	A
voltage, potential difference	V	volt	V
resistance	R	ohm	Ω
total resistance	R_T	ohm	Ω
supply voltage	V_S	volt	V
resistance of resistor	$R_1, R_2...$	ohm	Ω
voltage across resistor	$V_1, V_2...$	volt	V
number of turns in primary coil	n_p	-	-
number of turns in secondary coil	n_s	-	-
voltage across primary coil	V_p	volt	V
voltage across secondary coil	V_s	volt	V
current in primary circuit	I_p	ampere	A
current in secondary circuit	I_s	ampere	A
voltage gain	V_{gain}	-	-
output voltage	V_o	volt	V
input voltage	V_i	volt	V
power gain	P_{gain}	-	-
output power	P_o	watt	W
input power	P_i	watt	W

Physical Quantity	Symbol	Unit	Abbreviation
period	T	second	s
frequency	f	hertz	Hz
wavelength	λ	metre	m
power of a lens	P	dioptre	D
activity	A	becquerel	Bq
number of nuclei decaying	N	-	-
absorbed dose	D	gray	Gy
radiation weighting factor	w_R	-	-
equivalent dose	H	sievert	Sv
half-life	$t_{½}$	second	s

1.2 Preparing for the exam

You should refer to this section while you are studying each unit. Pay particular attention to the advice on how to answer the different types of exam questions.

What You Should Know

On the day of your Intermediate 2 Physics exam you need to:

- be aware of what the exam involves
- know how to answer multiple-choice questions
- present answers to numerical questions so that you gain as many marks as possible
- write clear, understandable and relevant descriptions, explanations and conclusions
- manage your time so that you complete the paper.

The Int2 Physics exam

The Int2 Physics exam lasts for **2 hours** and is out of **100 marks**.

Roughly one third of the marks are for questions on Mechanics and Heat, and another third for questions on Electricity and Electronics. The remaining marks are split evenly between the other two units.

At the start of the paper there are 20 multiple-choice questions each worth one mark. The remaining 80 marks require written responses of a few words, a few sentences, or numerical calculations.

Approximately **50 marks** are allocated to **knowledge and understanding** questions.

Key Points

Knowledge and understanding questions test your ability to:

- use quantities and units
- use relationships to solve straightforward numerical questions
- apply Physics principles in familiar situations
- describe familiar models, e.g. the nuclear model of the atom.

It is very important that you score as many marks as possible for the knowledge and understanding questions – this will give you a solid basis for passing the exam and achieving a good grade.

You will only be able to score these marks if you prepare well for the exam – so **study properly!**

Approximately **50 marks** are allocated to **problem solving** questions.

Key Points

Problem solving questions test your ability to:

- select and present **relevant** information
- solve numerical questions in a variety of contexts
- draw **valid** conclusions from information
- explain observations
- plan, design, and evaluate experimental procedures
- integrate skills across units.

About half of the problem solving questions will be set in reasonably familiar situations.

The remaining questions may be set in contexts that you have never seen before. These questions are generally the most difficult in the exam. **Do not panic when you find questions like this.** You can easily pass your Int2 without getting marks for these questions. You do need to score marks in these questions if you want to get a grade A.

Multiple-choice questions

Do not expect all of the multiple-choice questions to be easy. Some will be straightforward and some will be difficult.

In Int2 Physics there is **one and only one** correct answer to a multiple-choice question. There are **four wrong answers** to distract you. A good way to avoid being distracted is to cover the possible answers while you read the question.

If you can, work out an answer before looking at the options given in the question paper. When you have decided what you think the answer is then look at the possible answers.

If you are not sure, improve your chances by eliminating answers that you know are definitely wrong. If you are **completely** stuck, then guess.

Do not spend too long on the multiple-choice questions – between 20 and 25 minutes should leave you enough time to complete the rest of the paper.

Numerical questions

When you are doing numerical questions you should use the following steps – this will help you get as many marks as possible.

Hints and Tips

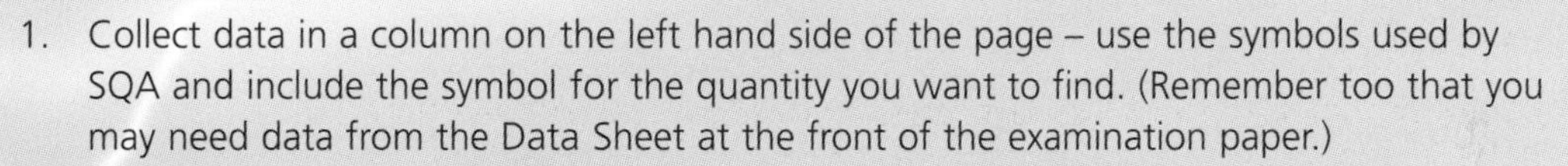

1. Collect data in a column on the left hand side of the page – use the symbols used by SQA and include the symbol for the quantity you want to find. (Remember too that you may need data from the Data Sheet at the front of the examination paper.)
2. Check the unit of each piece of data – make sure all have **correct SI units**.
3. Select the correct relationship from your Physics Data Booklet – look for a relationship that has the symbols included in the list you made at step 1.
4. Write the relationship in the centre of the page.
5. **On the line below** substitute one number at a time – **do not rearrange** before you substitute. Make sure the 'equals' signs are in line. (This makes it easier for you to check your working.)
6. Carry out the calculation – **do this at the right hand side of the page or on your calculator**.
7. Write down your answer in the form '*symbol* = number' – again make sure the 'equals' signs are in line.
8. If necessary, round to the correct number of significant figures (keep one extra figure in intermediate values).
9. Add the correct SI unit for the quantity you have calculated.

This method is illustrated in the examples which follow. (It is also used in all solutions to numerical questions in this book.)

Example

(How to present the answers to numerical problems.)

1 A cyclist moving at 3·1 m/s pedals faster for 8·0 seconds and reaches a speed of 4·3 m/s. Calculate her acceleration.

u = 3·1 m/s (*Collect the data in a column using symbols.*)

v = 4·3 m/s (***Check** that the units are correct SI units.*)

t = 8·0 s

a = ? (***Include** the symbol for the quantity you need to calculate.*)

(***Which relationship** includes these symbols?*)

$a = \frac{v-u}{t}$ (*Answer*)

$\Rightarrow a = \frac{4.3 - 3.1}{8.0}$ (***Substitute** carefully*)

$\Rightarrow a = 0.15\ \text{m/s}^2$. (*Do the **arithmetic**.*)

***Check** that the final answer has the correct **unit** and number of **significant figures**.*

2 The potential difference across a 4·8 kΩ resistor is 12 V. Calculate the current in the resistor.

R = 4·8 kΩ = 4800 Ω (*Collect data and check units.*)

V = 12 V

I = ? (*Which relationship has R, V and I?*)

$V = IR$ (*Answer*)

$\Rightarrow 12 = I \times 4800$ (*Substitute carefully.*)

$\Rightarrow I = 2.5 \times 10^{-3}$ A. (*Do the arithmetic.*)

(*Remember – significant figures and units!*)

Descriptions, explanations, and conclusions

In your Int2 Physics exam paper, the number of marks for each question, or part of a question, is shown in **bold** on the right hand side of the page.

The more marks there are – the more you usually have to write. Normally each **relevant** piece of Physics information is worth one mark. So to get full marks in a 2-mark question, you should include **two** relevant pieces of information.

In descriptions, explanations, and conclusions, the information you present must be **relevant, clear** and **complete**. After you have written an answer, read it to see if it makes sense.

In most explanation questions you need to **apply a Physics principle or relationship**. For example, questions about moving objects often involve Newton's laws. Electrical questions often involve current and voltage rules for series and parallel circuits.

Pay attention to verbs. 'Describe and explain' is not the same as 'describe' on its own. 'State' is easier to answer than 'justify'.

Use the language of Physics. There are many terms in Physics that have precise meanings – use these words in your written responses. Paraphrasing often introduces inaccuracy and may cost you marks.

Managing your time

This is a skill which you must practise – not just for Physics, but for all of your subjects. Every time you sit a class test, a prelim, or an exam, you can practise managing your time. Check your watch (or the clock in the exam room) every 15 minutes or so. You must make sure that you do not run out of time and that you try to answer all of the questions.

Chapter 2

MECHANICS AND HEAT

2.1 Speed, distance, and time

What You Should Know

For Intermediate 2 Physics you need to be able to:

- understand and use the equation $d = \bar{v}\, t$
- understand and use the terms 'average speed' and 'instantaneous speed'
- describe how to measure an average speed
- describe how to measure an instantaneous speed
- identify situations where instantaneous speed and average speed are different.

Average speed

Average speed is calculated by dividing the *total distance* by the *total time*.

The time measurement may include times when the object is stopped, when it is moving slowly, and when it is moving quickly.

The relationship between distance, average speed and time is usually written as

$$d = \bar{v}\, t \quad \textbf{or} \quad s = \bar{v}\, t \quad \text{where:}$$

d or s is **distance**, measured in *metres* (m)

$\bar{v}$ is **average speed**, measured in *metres per second* (m/s)

t is **time**, measured in *seconds* (s).

The bar above the symbol '*v*' means 'average'.

For an object moving at constant speed, its average speed equals its constant speed.

That is **for constant speed,** $\bar{v} = v$.

Instantaneous speed

The speed of an object at any instant is called the *instantaneous speed*. It is measured by calculating the average speed for a very short time around that instant. This is usually a very good estimate of the actual speed – the shorter the time used, the better the estimate.

For many types of motion the instantaneous speed changes frequently. For example a cyclist travelling to work in a city may have to start and stop because of traffic lights, traffic, and pedestrians. The cyclist has one average speed and many instantaneous speeds.

Question and Answer

A girl walks 0·96 km from her home to her school in a time of 12 minutes. Calculate her average speed on her way to school.

$\bar{v} = ?$ $\qquad d = \bar{v}\, t$

$d = 0{\cdot}96 \text{ km} = 960 \text{ m} \qquad \Rightarrow \quad 960 = \bar{v} \times 720$

$t = 12 \text{ min} = 720 \text{ s} \qquad \Rightarrow \quad \bar{v} = 1{\cdot}33$

$= 1{\cdot}3 \text{ m/s}.$

Measuring speed

Key Points

When you are asked to describe how to measure a speed you must:

- state both the **distance** and the **time** to be measured
- specify the **equipment** used to take measurements
- describe **how the measurements are taken**
- describe **how the measurements are used** to calculate speed.

Question and Answer

Describe how to measure the average speed of a trolley rolling down a ramp.

1. Place a trolley at the top of a ramp and mark its starting position. Put another mark at the bottom of the ramp. Use a metre stick to measure the distance, in metres, between the marks.
2. Release the trolley and, at the same time, start a stop clock. Stop the clock when the trolley reaches the second mark. Note the time reading in seconds. Repeat until at least five time readings are noted. Calculate the average time.
3. Divide the measured distance by the average time to find the average speed of the trolley in metres per second.

Exercises

Exercise 1 Speed, distance, and time

1 A boy walks to school in the morning and runs home from school in the afternoon. The distance between his home and the school is 360 m. The boy takes 180 s to walk to school and 120 s to run home. Calculate:

(a) the average speed of the boy on his way to school

(b) the average speed of the boy on his way home.

2 The average speed of a car is 25 m/s. Calculate the distance travelled by the car in 30 minutes.

3 The average speed of a yacht is 3·8 m/s. The total distance travelled by the yacht is 11·4 km. Calculate the total time for the yacht to travel this distance.

4 A girl is running 800 m as part of her training for a race. She wants to have an average speed of 5·0 m/s for the whole run. The girl runs the first 400 m in a time of 160 s. Explain whether she can reach an average speed of 5·0 m/s for the whole run.

5 A boy kicks a ball towards a target. A group of pupils is asked to measure the instantaneous speed of the ball just after it has been kicked. Another group of pupils is asked to measure the average speed of the ball from the moment it is kicked until it hits the target.

(a) Describe a method that the first group of pupils can use to measure the instantaneous speed of the ball.

(b) Describe a method that the second group of pupils can use to measure the average speed of the ball.

(c) Explain whether the instantaneous speed measured by the first group of pupils is likely to be greater than, equal to, or less than the average speed measured by the second group of pupils.

6 A racing car takes 0·080 s to cross the finishing line. The length of the car is 4·0 m.

(a) Calculate the speed of the car as it finishes the race.

(b) Explain whether this is an average speed or an instantaneous speed.

2.2 Scalars and vectors

What You Should Know

For Intermediate 2 Physics you need to be able to:

- define the terms scalar and vector
- state the difference between distance and displacement
- state the difference between speed and velocity
- define the term acceleration
- understand and use the terms 'speed', 'velocity' and 'acceleration'.

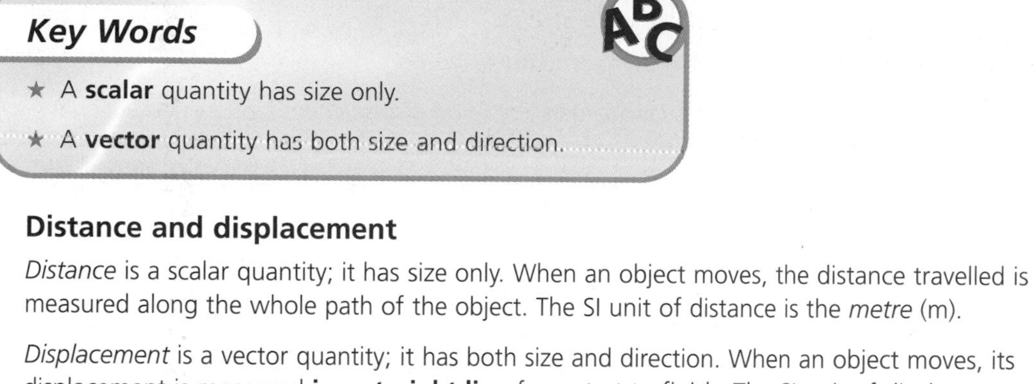

Key Words

★ A **scalar** quantity has size only.

★ A **vector** quantity has both size and direction.

Distance and displacement

Distance is a scalar quantity; it has size only. When an object moves, the distance travelled is measured along the whole path of the object. The SI unit of distance is the *metre* (m).

Displacement is a vector quantity; it has both size and direction. When an object moves, its displacement is measured **in a straight line** from start to finish. The SI unit of displacement is the *metre* (m).

In some numerical and graphical questions, *either* term 'distance' or 'displacement' may be used. This is appropriate for objects which:

- travel in a straight line, and
- do not change direction.

Question and Answer

An object starts at A and moves in straight lines first to B, then to C, then to D and finally to E where it stops.

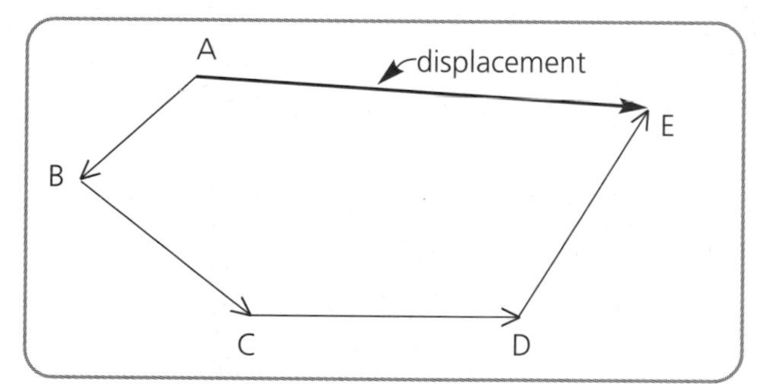

(a) How is the total distance moved by the object calculated?

(b) How is the total displacement of the object calculated?

(a) The total distance moved by the object is the sum of the lengths of lines AB, BC, CD, and DE. Hence $d = AB + BC + CD + DE$.

(b) The total displacement is represented by the line AE. The *size* of the displacement is the length of AE. Hence $d = AE$.

The direction of the displacement is the direction of line AE.

Speed and velocity

Speed is a scalar quantity. It is calculated by dividing *distance* by time. Speed has size only. The SI unit of speed is the *metre per second* (m/s).

Velocity is a vector quantity. It is calculated by dividing *displacement* by time. Velocity has both size and direction – the direction of a velocity is the same as the direction of the displacement used to calculate it. The SI unit of velocity is the *metre per second* (m/s).

In some numerical and graphical questions, *either* term 'speed' or 'velocity' may be used. This is appropriate for an object which travels in a straight line and does not change direction.

Acceleration

Acceleration is a vector quantity. It is defined as the *change in velocity* in unit time. The SI unit of acceleration is the *metre per second per second* (m/s^2).

Exercises

Exercise 2 Scalars and vectors

1 (a) Which of the following quantities are scalars?
- displacement
- speed
- acceleration
- velocity
- distance.

(b) Which of the above quantities are vectors?

2 State the difference between distance and displacement.

3 State the difference between velocity and speed.

4 (a) Which of the following quantities have direction?
- mass
- energy
- force
- work
- weight.

(b) Explain whether any of the above quantities are vectors.

5 A boy states that when an object is moving in a straight line, the distance travelled and displacement are always the same size. Is the boy correct? Give a reason for your answer.

2.3 Speed-time graphs and velocity-time graphs

What You Should Know

For Intermediate 2 Physics you need to be able to:

- describe motions shown by speed-time and velocity-time graphs
- draw speed time and velocity-time graphs involving more than one acceleration.

Speed-time graphs

Speed-time graphs are often used to represent the motion of objects, including those which are not moving in a straight line.

You **must** be able to recognise and draw the following three shapes of graph.

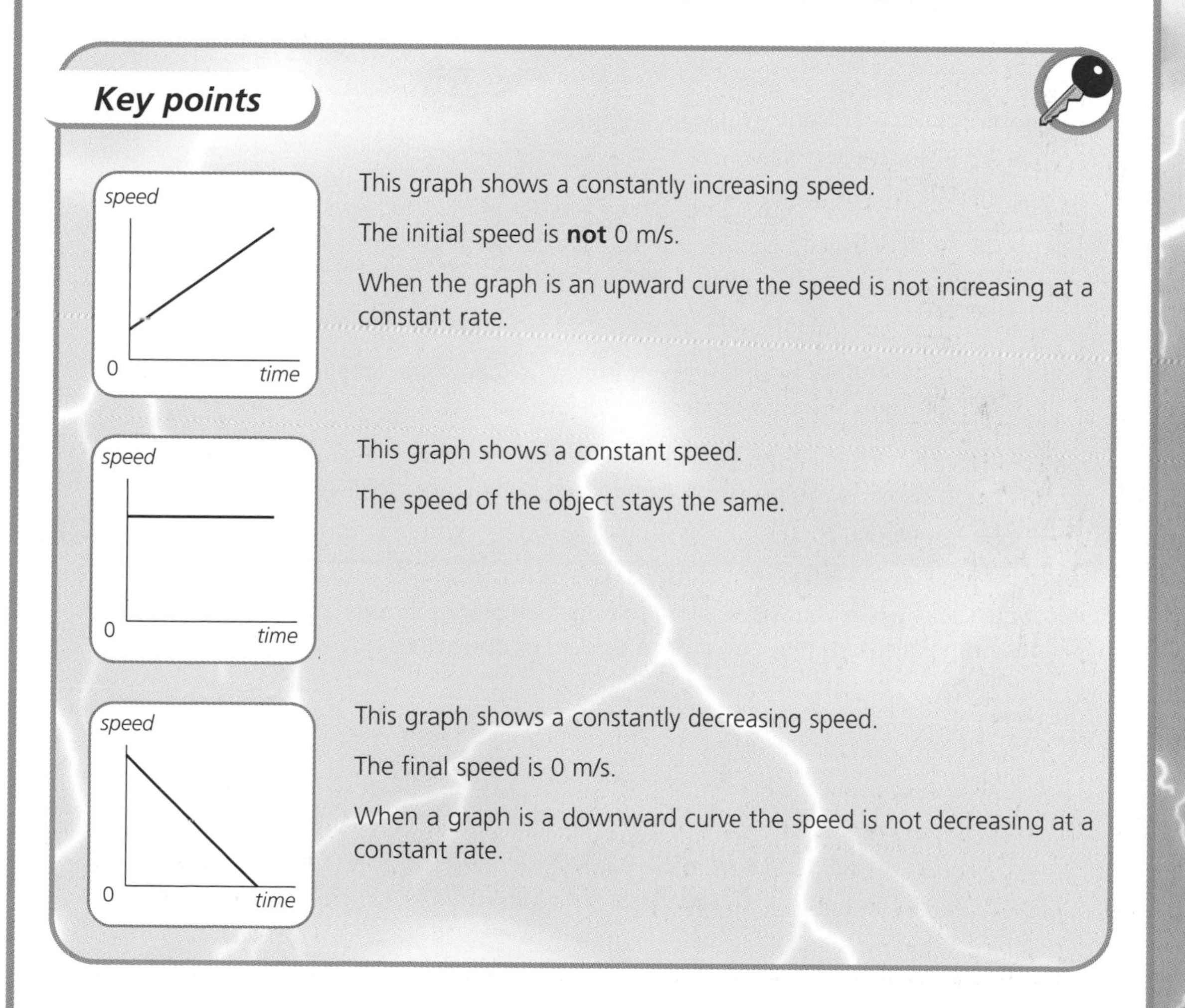

Key points

This graph shows a constantly increasing speed.

The initial speed is **not** 0 m/s.

When the graph is an upward curve the speed is not increasing at a constant rate.

This graph shows a constant speed.

The speed of the object stays the same.

This graph shows a constantly decreasing speed.

The final speed is 0 m/s.

When a graph is a downward curve the speed is not decreasing at a constant rate.

Most questions which you will meet on speed-time graphs include a combination of these shapes. Some questions may also include upward or downward curves.

Velocity-time graphs

The motion of an object moving in a straight line can be represented on a velocity-time graph. You must be able to recognise and draw the following three shapes of graph.

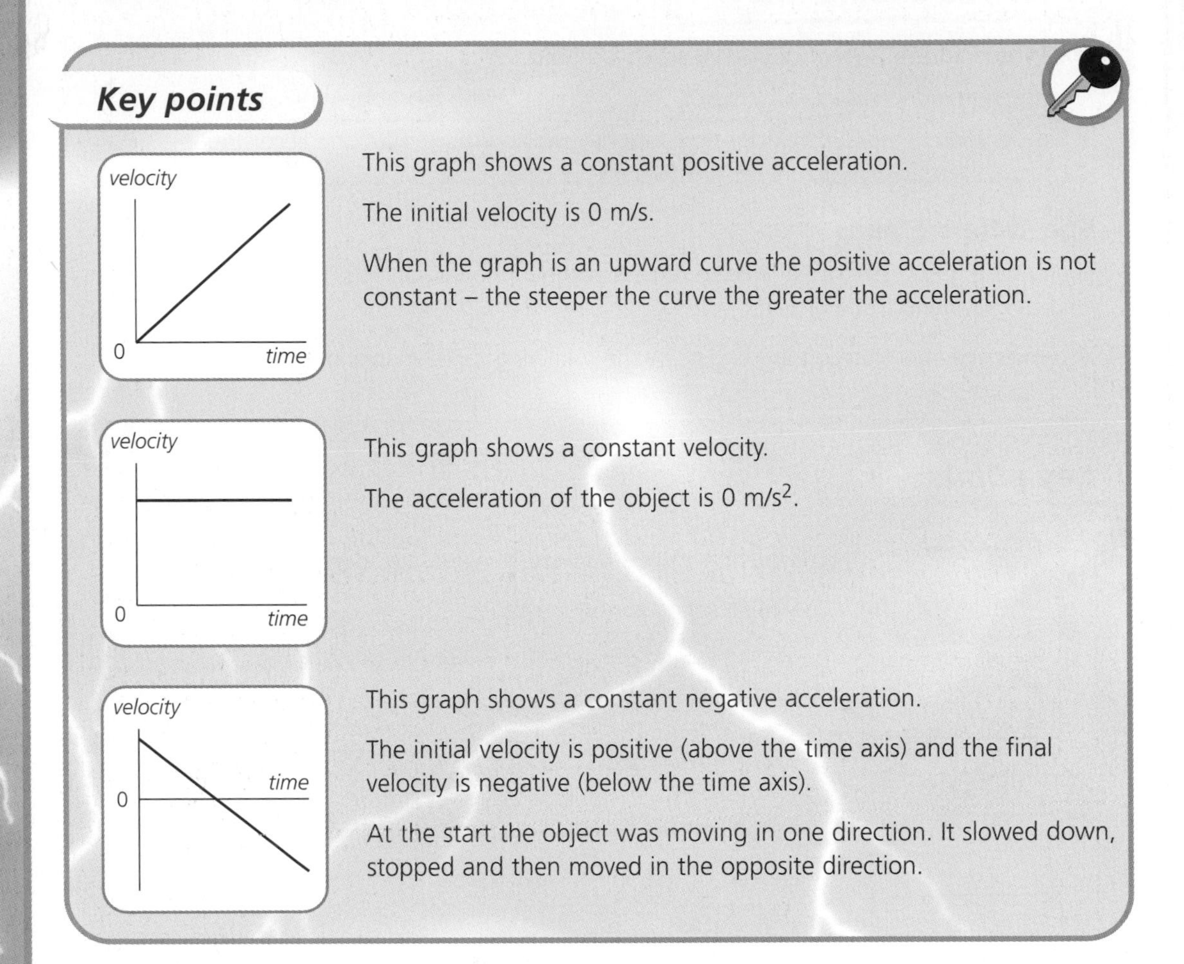

Key points

This graph shows a constant positive acceleration.

The initial velocity is 0 m/s.

When the graph is an upward curve the positive acceleration is not constant – the steeper the curve the greater the acceleration.

This graph shows a constant velocity.

The acceleration of the object is 0 m/s^2.

This graph shows a constant negative acceleration.

The initial velocity is positive (above the time axis) and the final velocity is negative (below the time axis).

At the start the object was moving in one direction. It slowed down, stopped and then moved in the opposite direction.

Most questions which you will meet on velocity-time graphs include a combination of these shapes. Some questions may also include upward or downward curves.

Exercises

Exercise 3 Speed-time graphs and velocity-time graphs

1 A speed-time graph is as shown. Look carefully at the graph and answer the questions which follow.

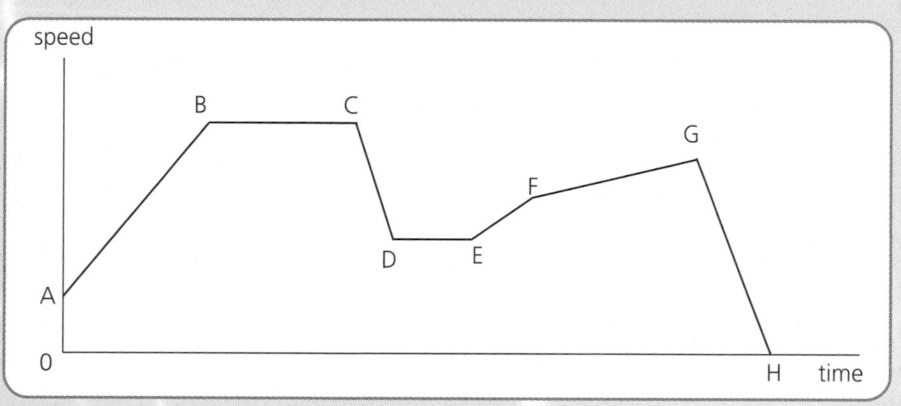

(a) Is the initial speed of the object 0 m/s? Explain your answer.

(b) Between which pair or pairs of points is the speed of the object:

(i) increasing, (ii) decreasing, (iii) constant ?

(c) State the final speed of the object.

2 The velocity-time graph of an object is as shown. Look carefully at the graph and answer the questions which follow.

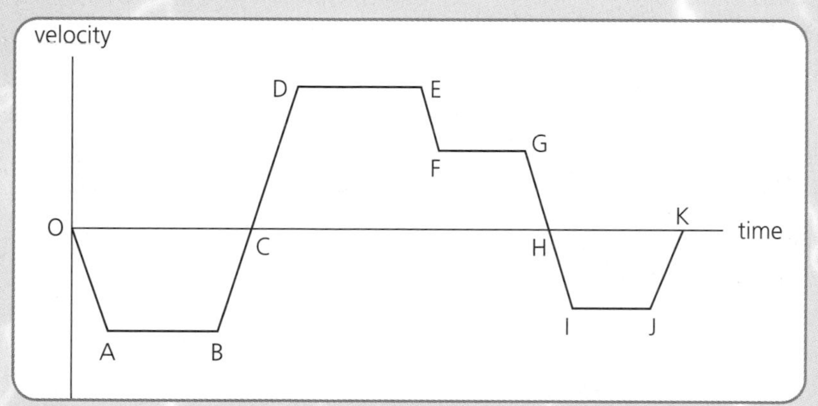

(a) Between which pair or pairs of points is the velocity of the object:

(i) positive, (ii) negative ?

(b) At which points is the velocity of the object 0 m/s?

(c) Between which pair or pairs of points is the **acceleration** of the object:

(i) positive, (ii) negative, (iii) zero ?

Exercises *continued* ➢

Exercises continued

3 A car, travelling on a long straight road, starts from rest and accelerates to a speed of 20 m/s in a time of 15 s. The car then travels at this speed for 45 s. The driver notices a friend walking along the roadside and applies the brakes for 20 s bringing the car to a stop some distance past his friend. The driver keeps the car stationary for 5s, then reverses back reaching a speed of 5 m/s after 5 s. He maintains this speed for 10 s, and then brings the car to rest in a further 5 s.

Draw the speed-time graph of the car.

2.4 Graphs, distance, displacement, and acceleration

What You Should Know

For Intermediate 2 Physics you need to be able to:

- calculate distance from a speed-time graph
- calculate displacement from a velocity-time graph
- calculate acceleration from a velocity-time graph
- understand and use relationships for calculating acceleration from velocity and time.

Speed-time graphs and distance

The *distance* travelled by an object is equal to the area between its *speed-time* graph and the time axis.

All areas must be counted and all areas are positive.

Question and Answer

The following graph represents the motion of a toy car.

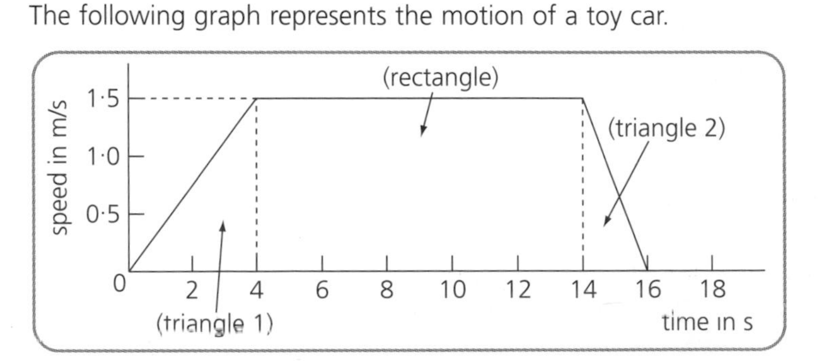

Use the graph to calculate the distance moved by the car.

***Question** and **Answer** continued* ➢

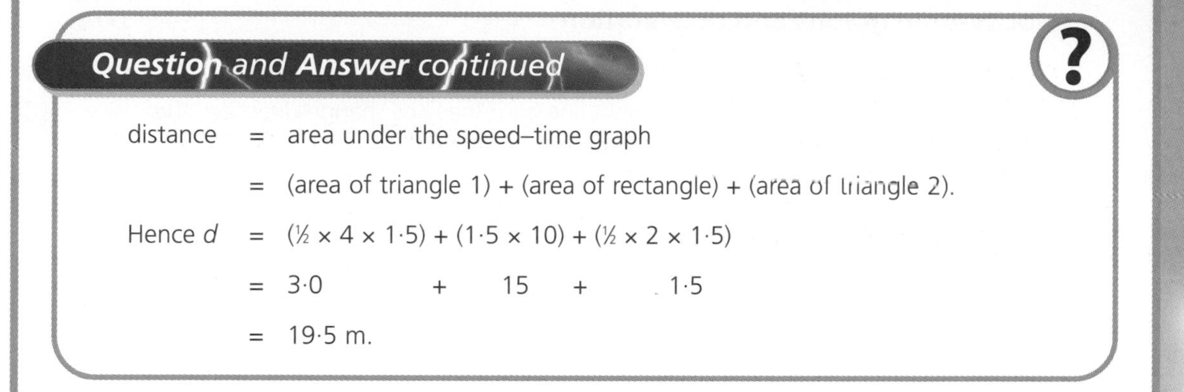

Question and Answer continued

distance = area under the speed–time graph

= (area of triangle 1) + (area of rectangle) + (area of triangle 2).

Hence d = (½ × 4 × 1·5) + (1·5 × 10) + (½ × 2 × 1·5)

= 3·0 + 15 + 1·5

= 19·5 m.

Velocity-time graphs and displacement

The *displacement* of an object is equal to the area between its *velocity-time* graph and the time axis. All areas must be counted.

Areas *above* the time axis are *positive*, and areas *below* the time axis are *negative*.

Question and Answer

The graph represents the motion of a ball thrown into the air.

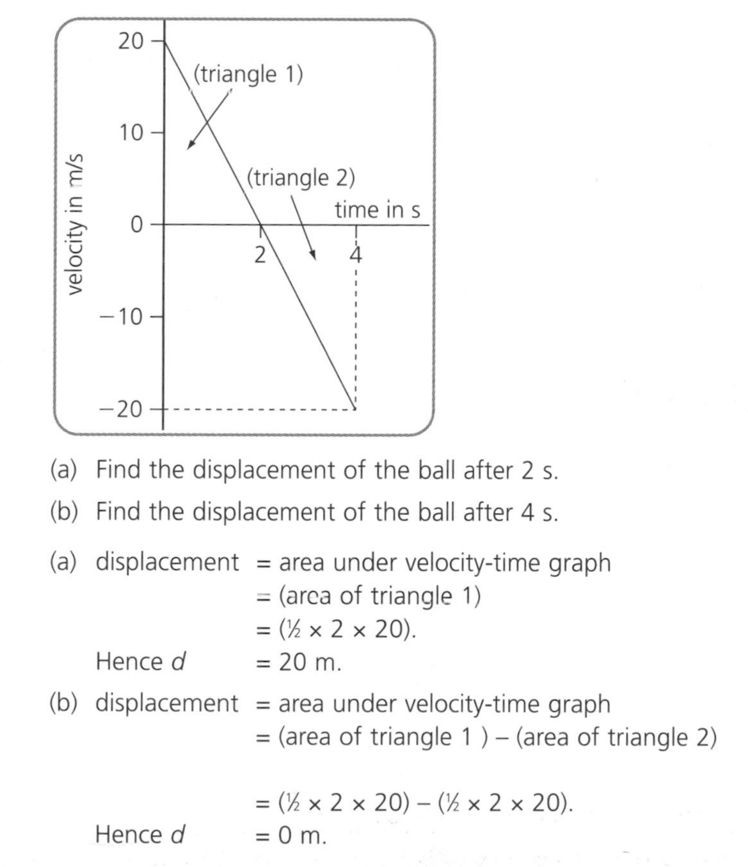

(a) Find the displacement of the ball after 2 s.

(b) Find the displacement of the ball after 4 s.

(a) displacement = area under velocity-time graph
= (area of triangle 1)
= (½ × 2 × 20).
Hence d = 20 m.

(b) displacement = area under velocity-time graph
= (area of triangle 1) – (area of triangle 2) (*triangle 2 is below the time axis*)
= (½ × 2 × 20) – (½ × 2 × 20).
Hence d = 0 m.

Velocity-time graphs and acceleration

The *acceleration* of an object is equal to the gradient of its *velocity-time* graph.

(For an object which is travelling in a straight line and does not change direction, the acceleration is equal to the gradient of its *speed-time* graph.)

Question and Answer

The graph represents the motion of a bus travelling between two bus stops.

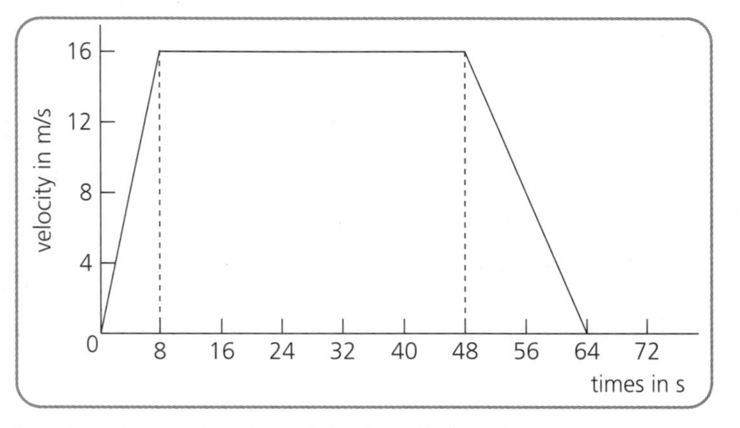

(a) Find the acceleration of the bus during the first 8 s.

(b) State the acceleration of the bus at time, t = 30 s. Give a reason for your answer.

(c) Find the acceleration of the bus at time, t = 60 s.

(a) acceleration = gradient of velocity-time graph

$$= \frac{16 - 0}{8 - 0}$$

Hence a = 2·0 m/s^2.

(b) At t = 30 s, a = 0 m/s^2. (The graph is parallel to the time axis and has zero gradient.)

(c) At t = 60 s, a = acceleration between t = 48 s and t = 64 s

= gradient of velocity-time graph

$$= \frac{0 - 16}{64 - 48}$$

= -1·0 m/s^2. (*this acceleration is negative*)

Calculating acceleration

The relationship for calculating acceleration is $a = \frac{\Delta v}{t}$. In this relationship,

a is **acceleration**, measured in *metres per second per second* (m/s^2)

Δv is **change in velocity**, measured in *metres per second* (m/s)

t is **time**, measured in *seconds* (s).

This relationship is also often written as $a = \frac{v - u}{t}$. In this form,

a is **acceleration**, measured in *metres per second per second* (m/s^2)

v is **final velocity**, measured in *metres per second* (m/s)

u is **initial velocity**, measured in *metres per second* (m/s)

t is **time**, measured in *seconds* (s).

Be very careful when you substitute values in this relationship. Always make sure that *v* and *u* are in the correct places – incorrect substitution of these values is **wrong physics**. You do not gain any marks after an incorrect substitution.

Exercises

Exercise 4 Graphs, distance, displacement, and acceleration

1 The graph represents the speed of a cyclist on a cycle track.

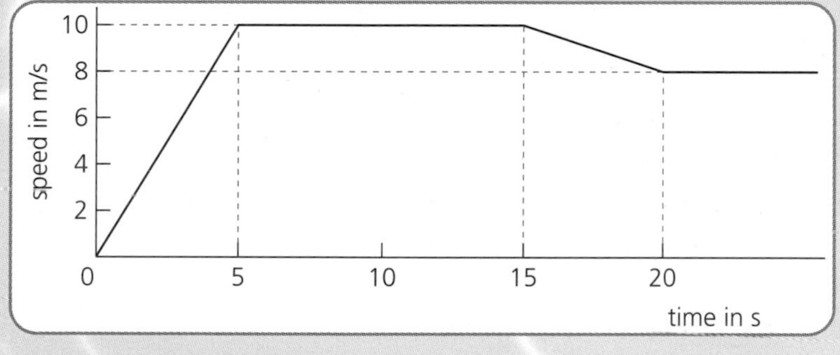

(a) Calculate the distance travelled by the cyclist in the first 5 s.

(b) Calculate the distance travelled by the cyclist in the first 15 s.

(c) (i) How far does the cyclist travel between t = 15 s and t = 20 s?

(ii) Give a reason for the cyclist slowing down between these times.

***Exercises** continued* ➢

Exercises continued

2 The graph represents the motion of a bouncing ball.

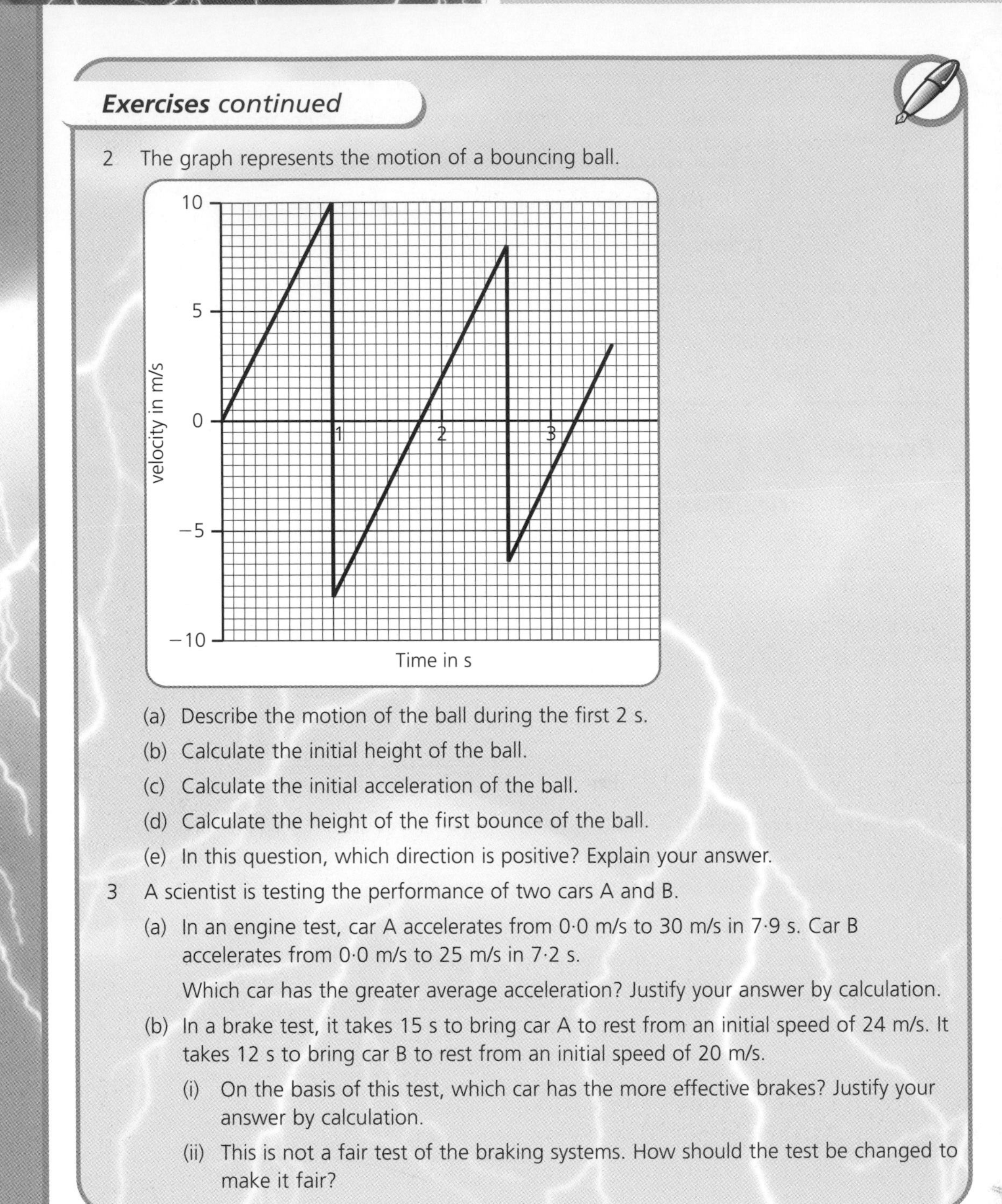

(a) Describe the motion of the ball during the first 2 s.

(b) Calculate the initial height of the ball.

(c) Calculate the initial acceleration of the ball.

(d) Calculate the height of the first bounce of the ball.

(e) In this question, which direction is positive? Explain your answer.

3 A scientist is testing the performance of two cars A and B.

(a) In an engine test, car A accelerates from 0·0 m/s to 30 m/s in 7·9 s. Car B accelerates from 0·0 m/s to 25 m/s in 7·2 s.

Which car has the greater average acceleration? Justify your answer by calculation.

(b) In a brake test, it takes 15 s to bring car A to rest from an initial speed of 24 m/s. It takes 12 s to bring car B to rest from an initial speed of 20 m/s.

(i) On the basis of this test, which car has the more effective brakes? Justify your answer by calculation.

(ii) This is not a fair test of the braking systems. How should the test be changed to make it fair?

2.5 Force

What You Should Know

For Intermediate 2 Physics you need to be able to:

- describe the effects of forces
- describe how to measure force using a newton balance
- understand and use the terms 'force', 'mass', 'weight' and 'gravitational field strength'
- carry out calculations using the equation $W = mg$
- state that the force of friction can oppose the motion of an object.

Force is a vector quantity.

The SI unit of force is the *newton*.

(A force of 1 newton is approximately equal to the weight of an average sized apple.)

Remember

A force can change:

- the shape of an object
- the speed of an object
- the direction of motion of an object.

These are the **effects** of forces. When you observe one of these effects you know that a force is acting.

Measuring force

In many experiments in Int2 Physics, force is measured using a newton balance. This is a balance in which the scale is marked in newtons. In the Int2 Physics exam you could be asked to describe how to use a newton balance, as in the following example.

Question and Answer

Describe how to use a newton balance to measure the force required to pull a block along the surface of a desk.

First attach the balance to the block, as shown.

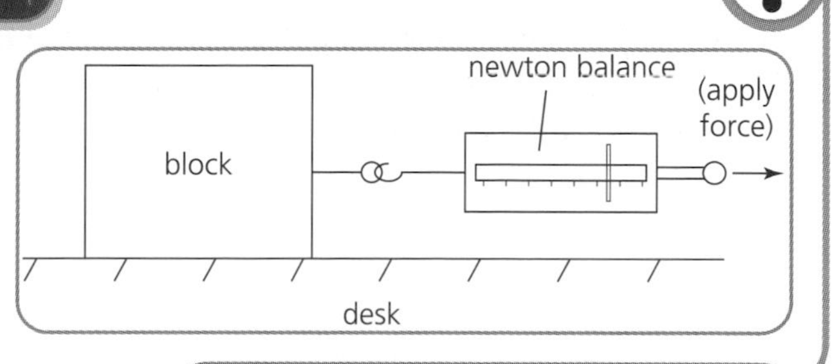

***Question** and **Answer** continued ➢*

Question and Answer continued

Then pull on the balance in the direction of the arrow. Try to ensure that the balance is horizontal. When the block is moving note the reading on the balance. Repeat until at least five measurements have been obtained and calculate the average.

Mass and weight

Mass is a scalar quantity.

The SI unit of mass is the *kilogram* and the symbol for mass used by SQA is m. The mass of an object does not change when it is moved – even if the object goes into space, or to another planet!

Weight is a force and so it is a vector quantity.

The SI unit of weight is the *newton* and the symbol for weight used by SQA is W.

On Earth, the weight of an object is the pull of planet Earth on the object. The weight of an object can and does change when it is moved. For example, the weight of an object on Mars is different from the weight of the object on Jupiter, or on Earth.

Gravitational field strength

The weight W of an object is calculated as the product *mass* × *gravitational field strength*.

The symbol used by SQA for gravitational field strength is g.

The SI unit of gravitational field strength is the *newton per kilogram* (N/kg).

$$\text{Thus} \quad W = mg.$$

On Earth the gravitational field strength is approximately 10 N/kg.

If you are asked a question about the weights of objects or people on say the moon, the Earth, or on any other planets, you will be given the gravitational field strengths in the question, or there will be a table of gravitational field strength values inside the front cover of the examination paper.

Gravitational field strength is numerically equal to the *acceleration due gravity*. The SQA uses the same symbol 'g' for both quantities. Though acceleration is measured in different units (*metres per second per second*), gravitational field strength and acceleration due to gravity are equivalent.

Gravitational field strength ≡ *acceleration due to gravity.*

(The symbol in the middle is **not** 'equals' – it has three lines and means 'is equivalent to'.)

Question and Answer

An astronaut has a weight of 620 N on Earth. Calculate her weight on the moon.

(Gravitational field strength of the moon = 1·6 N/kg.)

On Earth, $W = 620$ N, $g = 10$ N/kg, $m = ?$

$W = mg$

$\Rightarrow 620 = m \times 10$

$\Rightarrow m = 62$ kg.

On the moon, $m = 62$ kg, $g = 1{\cdot}6$ N/kg, $W = ?$

$W = mg$

$= 62 \times 1{\cdot}6$

$= 99{\cdot}2$

$= 99$ N.

(Why is the final answer rounded to 2 figures?) (Think – ***significant*** figures!)

The force of friction

Friction is a force which is very common in everyday life.

For example, think of the difference between the motion of a falling hammer and a falling feather!

Friction can oppose the motion of moving objects.

Sometimes the force of friction is helpful however, and so we may try to make it bigger.

For example, when a car is going around a corner, the force of friction between the car tyres and the road keeps the car on the road. A fully laden lorry going around the same corner and at the same speed as the car needs a much bigger force to keep it on the road, so its tyres are designed to make the force of friction large.

At other times the force of friction is not helpful and so we try to make it smaller. For example, a skier in a downhill ski race wants to cross the finish line as quickly as possible. The skier therefore tries to make the force of friction as small as possible. Can you think how this is achieved?

Exercises

Exercise 5 Force

1 (a) Describe two examples of a force changing the shape of an object.

(b) Describe two examples of a force changing the speed of an object.

(c) Describe two examples of a force changing the direction of motion of an object.

Exercises *continued* ➢

***Exercises** continued*

2 Describe how to use a newton balance to measure the weight of a pencil.

3 The mass of a car is 1200 kg. Calculate the weight of the car (on planet Earth).

4 A surface vehicle of mass $1{\cdot}8 \times 10^3$ kg is sent to collect samples of rocks from the surface of Mars.

Calculate the change in the weight of the surface vehicle as it travels from Earth to Mars.

(Gravitational field strength on Mars = 4 N/kg.
Gravitational field strength on Earth = 10 N/kg.)

5 (a) Describe one situation where the force of friction is not helpful. Describe one way of reducing this frictional force.

(b) Describe one situation where the force of friction is helpful. Describe one way of increasing this frictional force.

2.6 Force, resultant force, and motion

What You Should Know

For Intermediate 2 Physics you need to be able to:

- understand and use the terms 'balanced forces', 'unbalanced forces', and 'resultant force'
- explain the movement of an object in terms of the resultant force acting on the object
- use diagrams to analyse forces acting on an object
- find the resultant of two forces acting at right angles to each other
- define the newton
- carry out calculations using the equation $F = ma$.

Balanced forces and unbalanced forces

When you sit on a chair there are at least two forces acting on you – the force of gravity downwards, and the force from the chair upwards. These two forces are equal in size and opposite in direction – they cancel each other out. We say these forces are *balanced*.

Key Points

- ***When the forces acting on an object are balanced the motion of the object does not change***.

  A stationary object remains stationary. A moving object moves in a straight line at a constant speed. Balanced forces have the same effect on the motion of the object as no force at all.
- When the forces acting do not cancel each other, a *net force* acts on the object. We say that the forces are unbalanced. We call the net force the *resultant force*.
- ***When the forces acting on an object are unbalanced the motion of the object changes***.

  The object may get faster or slower or the direction of its motion may change.
- 'The motion of an object changes ***only*** if it is acted on by an unbalanced force.'

  This is **Newton's First Law**.

Question and Answer

For each of the following, explain whether the forces acting are balanced or unbalanced:

(a) a bus stationary at a bus stop

(b) a bus turning a corner.

(a) The bus is stationary – its motion is not changing so the forces are balanced.

(b) The bus is turning a corner – the direction of its motion is changing, so the forces acting on the bus are unbalanced

Analysing forces acting on objects

The first step is to name the forces and identify the directions in which they act.

For example, when an aircraft is flying there are 4 forces acting. (In your Int2 exam you are very unlikely to be asked a question more complicated than this.)

The diagram shows these forces and the directions in which they act.

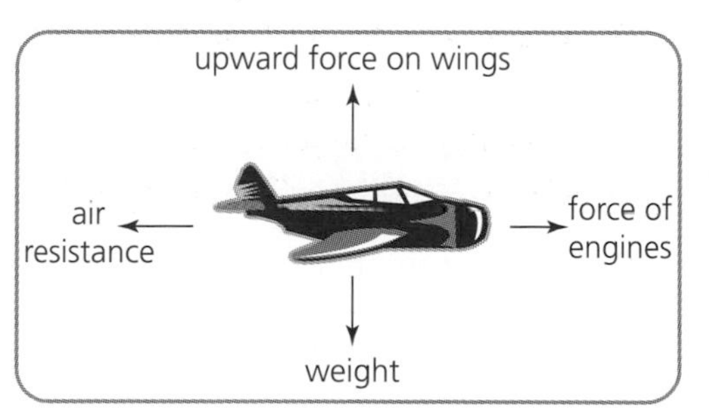

The second step is to consider the sizes of the forces acting on the object. (The motion of the object depends on the sizes of these forces and whether they are balanced or unbalanced.)

For example, when the aircraft is flying horizontally at a constant speed, the forces are balanced. The weight is balanced by the force on the wings; the force of the engines is balanced by air resistance. If the pilot then increases the engine force, the speed of the aircraft increases.

Question and Answer

A plane is flying horizontally at constant speed. The pilot reduces the engine force. Explain why the speed of the plane decreases.

When the plane is flying horizontally at a constant speed, the force of the engine is balanced by the force of air resistance – these two forces are the same size and they act in opposite directions.

When the force of the engine is reduced, it becomes smaller than the air resistance. The forces are now unbalanced. The unbalanced force then causes the plane to slow down.

Finding resultant force

For your Int2 Physics exam, you have to be able to find the resultant force acting on an object.

In some questions, the forces act along the same straight line. To find the resultant force you find the sum of these forces. **Remember** that force is a vector. Forces acting in opposite directions have opposite signs – one direction is positive, the other direction is negative.

Question and Answer

Three men are pushing a car up a slope. The amount of the weight of the car acting down the slope is 400 N. The men push up the slope with forces of 150 N, 120 N and 160 N. Calculate the resultant force acting on the car.

Total force up the slope = 150 + 120 + 160 = 430 N.

Total force down the slope = –400 N. (*Note this force is negative.*)

Hence resultant force = (430 – 400) = 30 N up the slope.

Forces at right angles

In some questions you have to find the resultant of two forces acting at right angles to each other. You can do this either by scale drawing or by using Pythagoras' Theorem.

Question and Answer

In an experiment, three students are using strings attached to newton balances to investigate the resultant of two forces. Students A and B use newton balances to apply forces of 3·6 N and 2·7 N at right angles to each other, as shown.

***Question** and **Answer** continued* ➢

Question and Answer continued

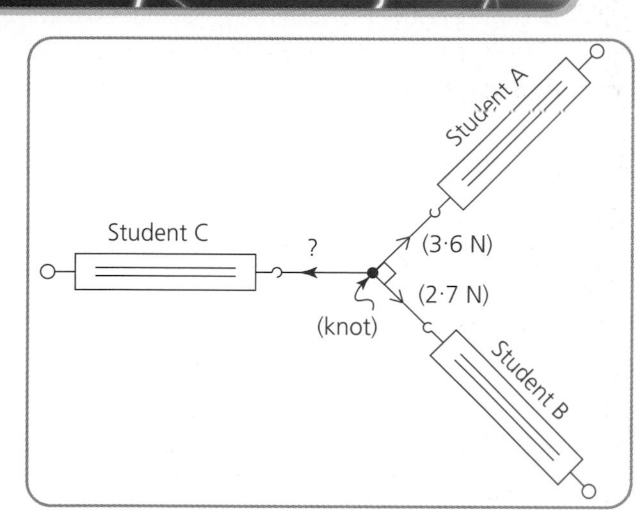

Student C holds a third newton balance steady.

(a) By using a scale drawing, find the resultant of the forces applied by students A and B.

(b) State the size of the reading on the newton balance held by student C. Explain your answer.

(a)

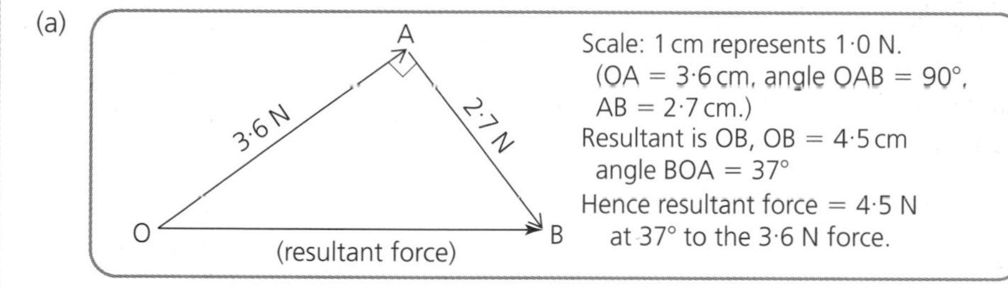

Scale: 1 cm represents 1·0 N.
(OA = 3·6 cm, angle OAB = 90°, AB = 2·7 cm.)
Resultant is OB, OB = 4·5 cm
angle BOA = 37°
Hence resultant force = 4·5 N at 37° to the 3·6 N force.

(b) Reading on newton balance held by student C = 4·5 N.

The force exerted by the balance held by student C is equal and opposite to the resultant of the two forces applied by students A and B.

Force mass and acceleration

When an unbalanced (resultant) force acts on an object, the object accelerates.

Key Words

The *newton* is defined as the resultant force which causes an acceleration of 1 m/s^2 when it acts on a mass of 1 kg. Learn this definition.

The relationship between resultant force, mass and acceleration is $F = ma$.

This relationship is the mathematical version of Newton's Second Law. You need to be able to use this relationship to solve numerical problems. Always remember that F stands for *resultant force*.

You should also use this relationship if you are asked to describe how changing mass or force affects the acceleration of an object.

Use the relationship in the form $a = \frac{F}{m}$, and consider how the value of $\frac{F}{m}$ changes when F or m changes.

For example, when force is increased and mass stays the same, the value of a increases. When mass is increased and force stays the same the value of a decreases.

In many Int2 questions you will be given the value of the resultant force. In other questions you will have to work out the value of the resultant force before you use the relationship. (In some questions, resultant force is simply called 'force'.)

Exercises

Exercise 6 Force, resultant force, and motion

1 For each of the following, explain whether the forces acing are balanced or unbalanced:
   (a) a pencil sitting on a desk
   (b) a pencil falling off of a desk
   (c) a pencil bouncing as it hits the floor.

2 (a) Give two examples of situations where the forces acting on an object are balanced.
   (b) Give two examples of situations where the forces acting on an object are unbalanced.

3 For each of the following, give the names and directions of the forces acting on the object.
   (a) A sports car accelerating at the beginning of a straight part of the track.
   (b) A family car slowing before it reaches a bend in the road.
   (c) (i) State whether the forces in part (a) are balanced or unbalanced.
       (ii) State whether the forces in part (b) are balanced or unbalanced.

4 A train is travelling along a straight horizontal track. The engine of the train exerts a forward force of 8300 N. A total frictional force of 3200 N acts on the train.
   (a) Calculate the resultant force acting on the train.
   (b) The total mass of the train is $3{\cdot}4 \times 10^4$ kg. Calculate the acceleration of the train.

5 Two forces of 500 N are applied as shown to drag a heavy crate across a factory floor.
   The crate moves at a constant speed of 0·1 m/s in the direction shown.
   Calculate the force of friction between the base of the crate and the factory floor.

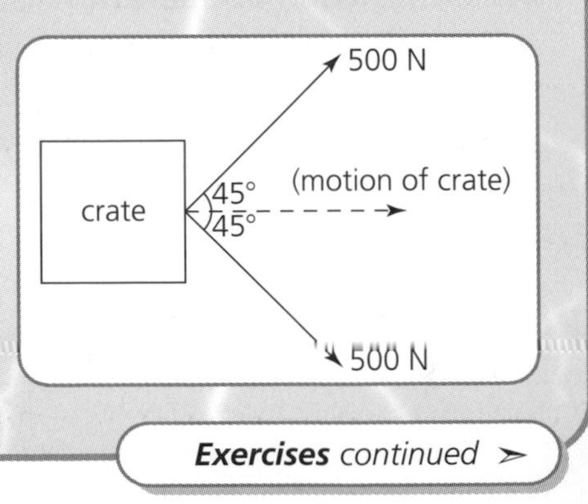

6 A car manufacturer is marketing a new model of a car but with an engine which produces a

Exercises continued ➢

Exercises continued

force 5% less than the standard model. The reduced power model is 10% cheaper. The mass of the reduced power model is 2% less than the standard car. How does the acceleration of the reduced power model compare with the acceleration of the standard model?

2.7 Projectiles

Projectiles are objects which are moving in an area of space where there is a gravitational force acting.

What You Should Know

For Intermediate 2 Physics you need to be able to:

- explain the shape of the path of a projectile
- explain that the movement of a projectile can be treated as two independent motions
- solve numerical problems on projectiles.

Vertical motion

When an object is dropped or thrown vertically upward, the motion of the object is along a single straight line. The acceleration of the object is constant. On Earth the vertical acceleration of such a projectile is 10 m/s^2.

In some questions the speed and shape of the projectile is small enough for air resistance to be ignored. In other questions however, air resistance has an effect on the motion of the object.

Question and Answer

A tennis ball is thrown vertically upwards with a velocity of 20 m/s. Calculate the time taken for the ball to return to its starting position. (Air resistance is negligible.)

$u = +20$ m/s

$a = -10$ m/s^2

$v = -20$ m/s

$t = ?$

$$a = \frac{v - u}{t}$$

$$\Rightarrow -10 = \frac{-20 - 20}{t} = \frac{-40}{t}$$ (*acceleration is negative since it is 'down'.*)

Hence $t = 4{\cdot}0$ s.

Horizontal and vertical motions

You may be asked a question about an object which is projected horizontally. Such an object has both vertical and horizontal motion.

The horizontal motion has no effect on the vertical motion. The vertical motion has no effect on the horizontal motion.

This is because these two motions are at right angles to each other.

Key Points

- When you are asked a question about an object projected horizontally, you must treat the two motions as if they are completely independent. The only thing which the two motions have in common is time.
- When air resistance can be ignored, the horizontal velocity of a projectile is constant.
- The vertical motion has a uniform acceleration downwards. It is exactly the same as for any object which is dropped.

Question and Answer

A football kicked horizontally from a vertical cliff has a vertical velocity of 30 m/s when it reaches the sea below.

(a) Calculate the time the ball takes to reach the sea.

(b) The initial horizontal velocity of the ball is 15 m/s. Calculate the horizontal distance travelled by the ball.

(a) $u = 0$ m/s — Using $a = \frac{v - u}{t}$ for vertical motion,

$v = 30$ m/s — $\Rightarrow \quad 10 = \frac{30 - 0}{t}$

$a = 10$ m/s^2 — $\Rightarrow \quad t = 3{\cdot}0$ s.

$t = ?$

(b) $v_{\text{horizontal}} = 15$ m/s — Using $d = vt$ (for constant horizontal velocity)

$t = 3{\cdot}0$ s — $\Rightarrow \quad d = 15 \times 3{\cdot}0$

$d = ?$ — $\Rightarrow \quad d = 45$ m.

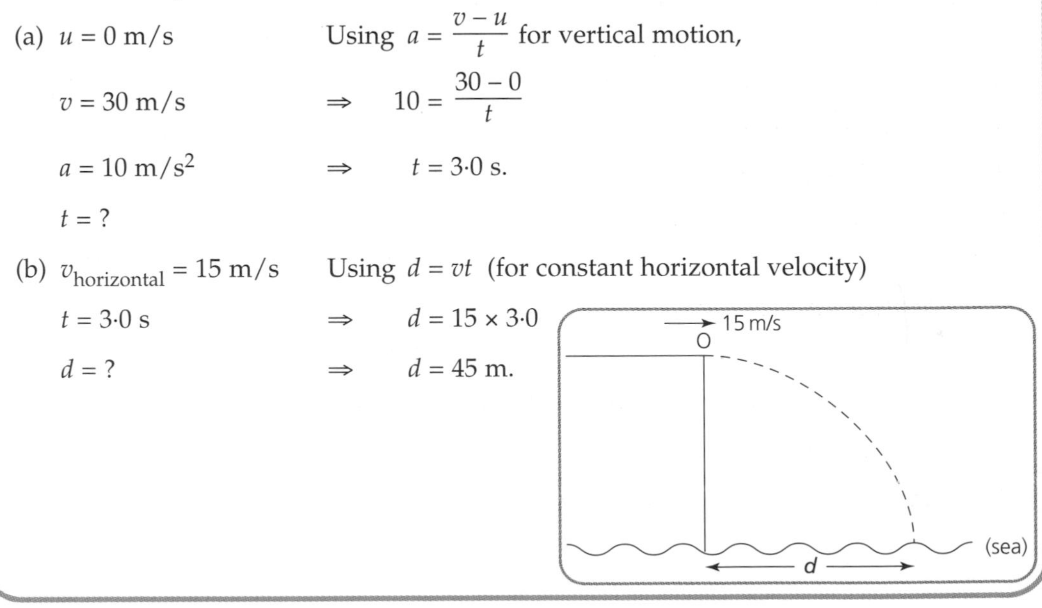

Exercises

Exercise 7 Projectiles

(In the questions which follow you may use the following: $g - 10$ m/s^2.)

1. A stone dropped down a well takes 3·0 s to reach the surface of the water. How far is the water surface below the top of the well?
2. A rocket fired vertically upward with a velocity of 60 m/s falls back to Earth. Ignoring the effects of air resistance, calculate the greatest height reached by the rocket.
3. An object which is projected horizontally follows a curved path. Explain the shape of the path.
4. A bird flying horizontally at 4·8 m/s drops a stone from its beak. The stone hits the ground after it has travelled a horizontal distance of 12 m.
   (a) After the bird dropped it, how long did it take the stone to fall to the ground?
   (b) Calculate the vertical velocity of the stone when it hits the ground.
5. At the start of a game of tennis, a girl hits a tennis ball so that its initial velocity is horizontal. The ball follows a curved path and bounces 1·5 s later.
   Using only this information and the value of *g*, is it possible to calculate the horizontal distance travelled by the tennis ball? Explain your answer.

2.8 Momentum and Newton's Third Law

What You Should Know

For Intermediate 2 Physics you need to be able to:

- understand and use the relationship $p = mv$
- understand and apply the law of conservation of linear momentum
- state Newton's Third Law
- identify forces which are 'Newton pairs'
- solve numerical problems on collisions in one dimension.

Momentum and its conservation

The *momentum* of an object is the product *mass* × *velocity*.

The symbol used by SQA for momentum is p. Hence $p = mv$.

Momentum is a vector quantity.

The SI unit of momentum is the *kilogram metre per second* (kg m/s).

Question and Answer

A trolley of mass 400 g has a velocity of 600 mm/s. Calculate the momentum of the trolley.

$p = ?$ $\qquad p = mv$

$m = 400\text{ g} = 0{\cdot}40\text{ kg}$ $\qquad \Rightarrow p = 0{\cdot}40 \times 0{\cdot}60$

$v = 600\text{ mm/s} = 0{\cdot}60\text{ m/s}$ $\qquad = 0{\cdot}24\text{ kg m/s}.$

Momentum is a very important quantity as physicists have discovered that in collisions, total momentum does not change, **provided there is no net external force acting**.

This is the *law of conservation of momentum*.

For objects moving in a straight line, the law is called the law of conservation of *linear* momentum.

Total momentum before collision = Total momentum after collision

You could be asked to state this law so learn it. The text in **bold** is very important. Many students sitting Int2 Physics forget to include this and lose marks as a result. Do not let this happen to you.

In the Int2 Physics exam you could be asked a question about a collision between two objects which move along the same straight line, and with one object initially stationary.

Question and Answer

Trolley A of mass 6·0 kg is rolling across a smooth horizontal desk with a velocity of 0·8 m/s. The trolley collides with a stationary trolley B of mass 2·0 kg. After the collision the trolleys couple and move off together in the direction in which A was travelling.

Calculate the velocity of the trolleys after the collision.

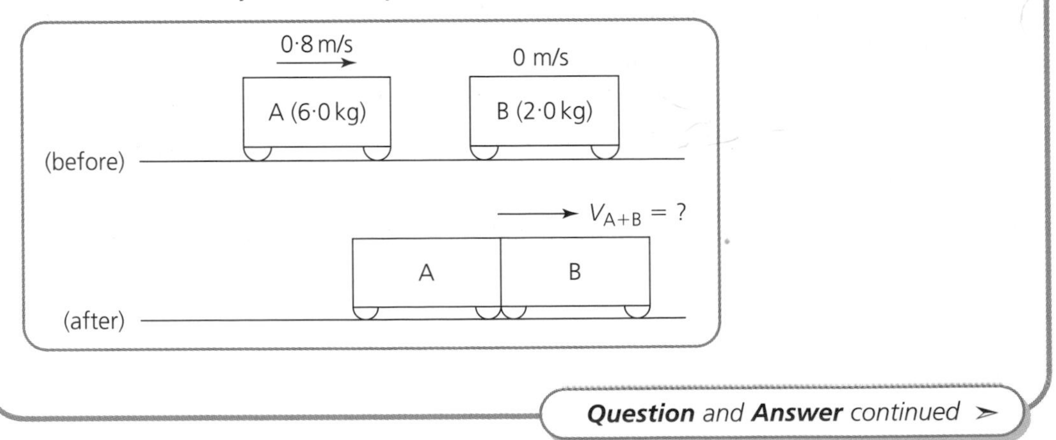

Question and **Answer** continued ➢

Question and Answer continued

Before the collision:

$p = ?$ Using $p = mv$

$m_{\text{trolley A}} = 6{\cdot}0\text{ kg}$ $p_{\text{trolley A}} = 6{\cdot}0 \times 0{\cdot}8$

$v_{\text{trolley A}} = 0{\cdot}8\text{ m/s}$ $= 4{\cdot}8\text{ kg m/s.}$

$p = ?$ Using $p = mv$

$m_{\text{trolley B}} = 2{\cdot}0\text{ kg}$ $p_{\text{trolley B}} = 6{\cdot}0 \times 0$

$v_{\text{trolley B}} = 0\text{ m/s}$ $= 0\text{ kg m/s.}$

Hence *total momentum* before collision = $4{\cdot}8\text{ kg m/s}$.

By the law of conservation of momentum, *total momentum* after collision = 4·8 kg m/s.

After the collision:

Using $p = mv$

$m_{\text{trolley A+B}} = 6{\cdot}0 + 2{\cdot}0 = 8{\cdot}0\text{ kg}$ $4{\cdot}8 = 8{\cdot}0 \times v_{\text{trolleys A+B}}$

$p_{\text{trolley A+B}} = 4{\cdot}8\text{ kg m/s}$ $\Rightarrow v_{\text{trolleys A+B}} = \frac{4{\cdot}8}{8{\cdot}0} = 0{\cdot}6\text{ m/s}$

$v_{\text{trolleys A+B}} = ?$ The trolleys move at 0·6 m/s after the collision.

Newton's Third Law

Newton's Third Law states 'Every force causes an equal and opposite force'.

When two objects interact, the force on each of them is the same. The two forces act in opposite directions.

For example, a girl standing on a floor exerts a downward force *on the floor*. The floor exerts an equal and opposite upward force *on the girl*. These two forces are called a 'Newton pair'.

In your Int2 exam you could be asked to identify forces which are Newton pairs.

Newton pair forces act between two objects. One force acts on one object, and the other acts on the other object.

Make sure that there are only two objects involved when you name Newton pair forces.

Question and Answer

For each of the following forces identify its Newton pair force.

(a) The force exerted by a spring on a trolley.

(b) The air resistance acting on an aeroplane in flight.

(c) The weight of a car.

The Newton pair forces are:

(a) The force exerted by the trolley on the spring.

(b) The force exerted by the aeroplane on the air.

(c) (Weight of a car – careful with this one!)

Weight is the force of gravity exerted by the Earth on the car – the Newton pair force is the force exerted by the car on the Earth.

Exercises

Exercise 8 Momentum and Newton's Third Law

1 A car of mass 1200 kg travels along a straight horizontal road with a constant speed of 15 m/s. Calculate the momentum of the car.

2 During a game of bowls, bowl A of mass 1·6 kg collides head-on with a stationary bowl B of equal mass. Immediately after the collision, bowl A has a speed of 0·30 m/s and bowl B has a speed of 0·60 m/s. Calculate the speed of bowl A immediately before the collision.

3 For each of the following forces identify its Newton pair force.

(a) The force exerted by a ship's propeller on sea water.

(b) The frictional force exerted on the shoe of a man walking.

(c) The force on a rocket in space.

4 Newton pair forces are equal in size and opposite in direction and yet they do not cancel each other out. Explain.

2.9 Work, energy, and power

What You Should Know

For Intermediate 2 Physics you need to be able to:

- state that work done is a measure of energy converted from one form to another
- understand and use the relationship $E_W = Fd$

What you should know *continued* ➢

What You Should Know continued

- understand and use the relationship $P = \frac{E}{t}$
- understand and use the relationship $E_p = mgh$
- understand and use the relationship $E_k = \frac{1}{2}mv^2$
- solve numerical problems on energy, work, power, and efficiency.

Conservation of energy

Energy cannot be created and it cannot be destroyed.

When an object loses energy, the lost energy is converted to other forms of energy.

When an object gains energy, the energy gained has been converted from other forms of energy.

This is true for all energy changes.

Work

When work is done, energy is converted from one form to another. The SI unit of work is the *joule* (J) – this is also the SI unit for energy.

The relationship for calculating work done when a force moves is $E_W = Fd$ where:

E_W is **work done**, measured in *joules* (J)

F is **force**, measured in *newtons* (N)

d is **distance moved by the force**, measured in *metres* (m).

One joule of work is done when a force of 1 N moves through a distance of 1 m.

Work is a scalar quantity.

Power

Power is the *rate* at which work is done, or energy is converted from one form to another.

The SI unit of power is the *watt* (W) and 1 watt is equal to 1 *joule per second* (J/s).

The relationship for calculating power is $P = \frac{E}{t}$ where:

P is **power**, measured in *watts* (W)

E is **energy**, measured in *joules* (J)

t is **time**, measured in *seconds* (s).

Question and Answer

A man exerts a force of 200 N for 6 minutes pushing his car 60 m along a horizontal road to reach a garage.

(a) Calculate the work done by the man.

(b) Calculate the power of the man when he was pushing the car.

(a) $E_W = ?$ $\qquad E_W = Fd$

$F = 200\text{ N}$ $\qquad \Rightarrow E_W = 200 \times 60$

$d = 60\text{ m}$ $\qquad = 12\,000\text{ J.}$

(b) $P = ?$ $\qquad P = \frac{E}{t}$

$E = 12\,000\text{ J}$ $\qquad \Rightarrow P = \frac{12\,000}{360}$

$t = 6\text{ mins} = 360\text{ s}$ $\qquad = 33{\cdot}3$

$= 33\text{ W.}$

Gravitational potential energy

Gravitational potential energy is energy gained or lost by an object as it moves vertically.

The relationship for calculating gravitational potential energy is $E_p = mgh$ where:

E_p is **gravitational potential energy**, measured in *joules* (J)

m is **mass**, measured in *kilograms* (kg)

g is **gravitational field strength**, measured in *newtons per kilogram* (N/kg)

h is change of **vertical height**, measured in *metres* (m).

In many Int2 questions, 'gravitational potential energy' is shortened to 'potential energy'. Gravitational potential energy is a scalar quantity.

Kinetic energy

Kinetic energy is energy which an object has because of its motion.

The relationship for calculating kinetic energy is $E_k = {}^1\!/_2 mv^2$ where:

E_k is **kinetic energy**, measured in *joules* (J)

m is **mass**, measured in *kilograms* (kg)

v is **velocity** (or speed) measured in *metres per second* (m/s).

Kinetic energy is a scalar quantity.

Question and Answer

A boy of mass 42 kg is standing at a height of 2·1 m above the ground on a climbing frame. The boy jumps to the ground.

(a) Calculate the potential energy lost by the boy during his jump.

(b) Hence calculate the speed of the boy when he lands.

(c) State any assumption you make to solve part (b).

(a) $E_p = ?$ $\quad$ $E_p = mgh$

$m = 42$ kg $\quad \Rightarrow \quad E_p = 42 \times 10 \times 2{\cdot}1$

$h = 2{\cdot}1$ m $\quad = 882$ J

$g = 10$ N/kg $\quad = 880$ J.

(b) $v =$ $\quad$ $E_k = \frac{1}{2}mv^2$

$E_k = 882$ J $\quad \Rightarrow \quad 882 = \frac{1}{2} \times 42 \times v^2$

$m = 42$ kg $\quad \Rightarrow \quad v = 6{\cdot}48$

$= 6{\cdot}5$ m/s.

(c) Assumption – all of the potential energy is converted to kinetic energy.

Efficiency

The *efficiency* of a machine or a process is a measure of how well the **useful** energy change takes place.

One relationship for calculating efficiency is *percentage efficiency* = $\frac{useful\ E_o}{E_i} \times 100$ where:

E_o is **energy out**, measured in *joules* (J)

E_i is **energy in**, measured in *joules* (J).

Efficiency has no unit – you will lose marks if you include a unit in the statement of your final answer when you calculate efficiency.

Efficiency may also be calculated from power using the relationship

percentage efficiency = $\frac{useful\ P_o}{P_i} \times 100$ where:

P_o is **power out**, measured in *watts* (W)

P_i is **power in**, measured in *watts* (W).

Question and Answer

The electric motor of a crane uses 42 000 J of electric energy lifting a pack of eight 25 kg bags of cement through a distance of 15 m from the ground to the fourth floor of a block of flats. Calculate the efficiency of the motor during the lifting process.

useful $E_o = E_p = ?$

$m = 8 \times 25 = 200$ kg

$g = 10$ N/kg

$h = 15$ m

$E_p = mgh$

$\Rightarrow E_p = 200 \times 10 \times 15$

$= 30\ 000$ J

$\Rightarrow$ useful $E_o = 30\ 000$ J.

%age efficiency = ?

$E_i = 42\ 000$ J

$$\%\text{age efficiency} = \frac{\text{useful } E_o}{E_i} \times 100$$

$$= \frac{30\ 000}{42\ 000} \times 100$$

$$= 71\%.$$

Exercises

Exercise 9 Work, energy, and power

1. A steady force of 30 N is used to move a small crate across a factory floor. The energy used moving the crate is 450 J. Calculate the distance moved by the crate.
2. A pendulum consists of a small metal sphere suspended at the end of a long string. The metal sphere is pulled to the side and then released from a point 35 mm higher than the lowest part of its swing. Calculate the maximum velocity of the metal sphere.
3. A boat travels at a constant speed of 6·0 m/s for 15 minutes. The input power of the boat engine is 12 000 W. The efficiency of the engine is 30%.
   (a) Calculate the energy used by the boat engine.
   (b) Calculate the useful energy output of the engine.
   (c) Calculate the force exerted by the engine.
4. Percentage efficiency has no unit. Explain why.

2.10 Heat and temperature

What You Should Know

For Intermediate 2 Physics you need to be able to:

- understand and use the terms 'heat', 'temperature', 'specific heat capacity' and 'specific latent heat'
- understand and use the relationship $E_h = cm\Delta T$
- understand and use the relationship $E_h = ml$

Heat and temperature

Heat is a form of energy.

The symbol used by SQA for heat is E_h and the SI unit for heat is the *joule* (J).

Temperature is a measure of how hot or how cold an object is.

The SI unit for temperature is the *degree celsius* (°C) and the symbol used by SQA for temperature is T.

Changing temperature

When an object gets hotter it gains heat energy.

When an object gets colder it loses heat energy. The amount of heat energy gained or lost depends on the **mass**, the **change in temperature** and the **material** of the object.

The heat energy gained when an object is heated is equal to the energy lost when it cools back down to its original temperature.

The relationship for calculating heat energy gained or lost by an object changing temperature is

$$E_h = cm\Delta T \quad \text{where:}$$

E_h is **heat energy**, measured in *joules* (J)

c is **specific heat capacity**, measured in *joules per kilogram per degree celsius* (J/kg °C)

m is **mass**, measured in *kilograms* (kg)

ΔT is **change of temperature**, measured in *degrees celsius* (°C).

Question and Answer

A block of copper of mass 0·33 kg is heated from 20 °C to 58 °C. Calculate the heat energy absorbed by the copper.

(Specific heat capacity of copper = 386 J/kg °C.)

$m = 0{\cdot}33$ kg

$\Delta T = 58 - 20 = 38$ °C

$c = 386$ J/kg °C

$E_h = ?$

$E_h = cm\Delta T$

$\Rightarrow E_h = 386 \times 0{\cdot}33 \times 38$

$= 4840{\cdot}44$

$= 4800$ J.

Changing state

When a solid object melts it gains heat energy. When the liquid turns to a solid it loses heat energy.

The amount of heat energy gained or lost depends on the **mass** and the **material** of the object.

The energy gained when a solid melts is equal to the heat energy lost when the liquid turns back into a solid.

In some Int2 questions the term 'fusion' is used to mean melting.

When a liquid turns to a gas it gains heat energy. When the gas turns to a liquid it loses heat energy.

The amount of heat energy gained or lost depends on the **mass** and the **material**.

The energy gained when a liquid turns to a gas is equal to the heat energy lost when the gas turns back into a liquid.

The relationship for calculating the heat energy gained or lost during a *change of state* is

$$E_h = ml \quad \text{where:}$$

E_h is **heat energy**, measured in *joules* (J)

m is **mass**, measured in *kilograms* (kg)

l is **specific latent heat**, measured in *joules per kilogram* (J/kg).

You must always keep change of state and change of temperature calculations separate – the change of state occurs at a constant temperature.

Question and Answer

The amount of heat energy required to melt completely a block of iron at its melting point is $3{\cdot}2 \times 10^6$ J. Calculate the mass of the iron block

(Specific latent heat of fusion (melting) of iron = $2{\cdot}67 \times 10^5$ J/kg.)

$E_h = 3{\cdot}2 \times 10^6$ J

$l = 2{\cdot}67 \times 10^5$ J/kg

$m = ?$

$$E_h = ml$$

$$\Rightarrow 3{\cdot}2 \times 10^6 = m \times 2{\cdot}67 \times 10^5$$

$$\Rightarrow m = \frac{3{\cdot}2 \times 10^6}{2{\cdot}67 \times 10^5}$$

$$= 11{\cdot}98$$

$$= 12 \text{ kg.}$$

Exercises

Exercise 10 Heat and temperature

1 Calculate the energy released when 3·1 kg of steam at 100 °C turns to water at 100 °C.
Specific latent heat of vaporisation of water = $2{\cdot}25 \times 10^6$ J/kg.

2 A block of ice of mass of mass 0·72 kg is at a temperature of –20 °C when it is removed from a freezer in a laboratory. The temperature of the air in the laboratory is 15 °C. The block of ice is placed in a container and allowed to melt. The melt water is then allowed to heat up to room temperature.

Specific heat capacity of ice = 2100 J/kg °C

Specific latent heat of fusion of ice = $3{\cdot}34 \times 10^5$ J/kg

Specific heat capacity of water = 4180 J/kg °C.

(a) Calculate the total energy absorbed by the ice and its melt water from leaving the freezer until reaching room temperature.

(b) State any assumption you have made in solving part (a).

Chapter 3

ELECTRICITY AND ELECTRONICS

3.1 Circuit basics

What You Should Know

For Intermediate 2 Physics you need to be able to:

- distinguish between and give examples of conductors and insulators
- understand and use the term 'current'
- understand and use the equation $Q = It$
- draw and identify circuit symbols
- draw circuit diagrams showing ammeters and voltmeters in the correct positions.

Conductors and insulators

A *conductor* is a material in which charges, usually electrons, are free to move.

Metals like iron, copper, brass, and aluminium are good conductors. Graphite (carbon) and water are also good conductors.

An *insulator* is a material in which charges are not free to move. Charge does not flow in insulators.

Glass, polythene and rubber are good insulators.

Charge flows when there is a complete path of conductors and a source of electrical energy. The complete path of conductors is called an *electrical circuit*.

Key Words and Definitions

A flow of charge in a conductor is called *current*.

The SI unit of current is the *ampere* and the symbol used by SQA is I.

The symbol used by SQA for charge is Q, and the SI unit of charge is the *coulomb*.

Charge, current, and time

For Int2 Physics you need to be able to use the relationship

$$Q = It \quad \text{where:}$$

Q is **charge**, measured in *coulombs* (C)

I is **current**, measured in *amperes* (A)

t is **time**, measured in *seconds* (s).

Question and Answer

A torch is switched on for 30 minutes. The current in the bulb of the torch is 0·45 A. Calculate the charge which flowed through the bulb.

$Q = ?$ $\qquad$ $Q = It$

$I = 0{\cdot}45$ A $\quad \Rightarrow \quad Q = 0{\cdot}45 \times 1800$

$t = 30$ min $= 1800$ s $\qquad = 810$ C.

Circuit symbols

You need to be able to draw and identify all of the symbols shown in the following table.

Component	Symbol	Component	Symbol
ammeter	—(A)—	voltmeter	—(V)—
battery	─┤├┄┤├─	resistor	─▭─
fuse	─▭─	variable resistor	─▭─ (with arrow)
switch	─/ ─	lamp	—(⊗)—

Series circuits

In a *series circuit* there is one and only one complete path of conductors. All of the components in the circuit are connected in one single loop.

The circuit here shows a battery, switch, ammeter, resistor, and lamp connected in series.

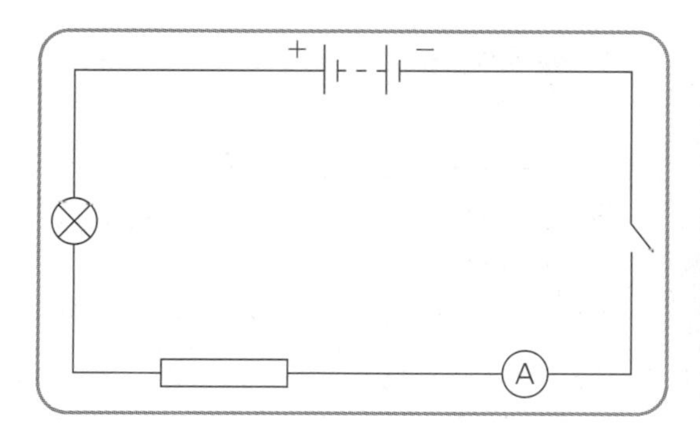

When you are using an ammeter to measure the current in an electrical component, the ***ammeter must always be connected in series*** with the component.

Parallel circuits

In a *parallel circuit* there is more than one complete path of conductors.

The following diagram shows a battery and a switch in series, two resistors in parallel, and a voltmeter connected to measure the voltage across the resistors.

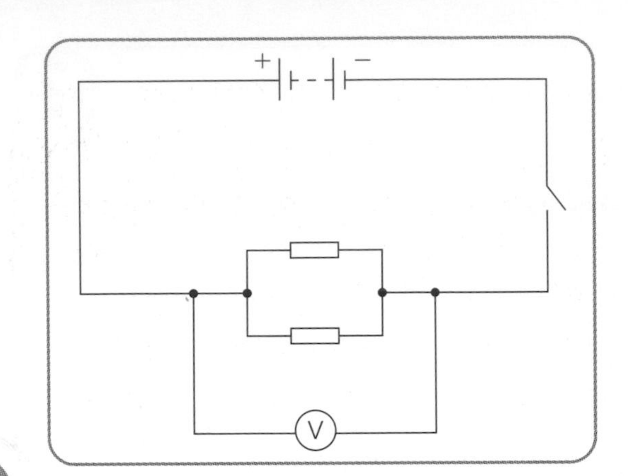

When you are using a voltmeter to measure the voltage across an electrical component, the ***voltmeter must always be connected in parallel*** with the component.

Exercises

Exercise 11 Circuit basics

1 State the two conditions which are necessary for charge to flow in a circuit.
2 Explain why both conductors and insulators are necessary for the operation of electrical circuits.
3 During a flash of lightning, 600 C of charge is transferred in 0·15 s. Calculate the average current.
4 Draw a series circuit containing a battery, switch, lamp, a variable resistor, and an ammeter connected so that it can be used to measure the current in the lamp.
5 Draw a circuit with a battery and switch in series, two lamps in parallel, and a voltmeter connected so that it can be used to measure the voltage across one of the lamps.

3.2 Current, voltage, and resistance

What You Should Know

For Intermediate 2 Physics you need to be able to:

- understand and use the terms 'voltage', 'resistance' and 'resistor'
- state the rules for current and voltage in series circuits
- state the rules for current and voltage in parallel circuits
- understand and use the relationship $V = IR$
- carry out calculations on series and parallel circuits.

Voltage and resistance

Key Words and Definitions

The **voltage** of an energy supply is a measure of the energy given to charges by the supply. The higher the voltage, the more energy the charges are given.

The SI unit of voltage is the *volt* and the symbol used by SQA is *V*.

The voltage across electrical components is sometimes called *potential difference* (p.d.).

Resistance is a measure of the opposition of a circuit component to the flow of charge.

The SI unit of resistance is the *ohm*, and the symbol used by SQA for resistance is *R*.

Increasing the resistance in a circuit causes the current to decrease.

A **resistor** is an electrical component whose resistance remains approximately constant for different currents.

For Int2 Physics you need to be able to use the following relationship which connects voltage, current and resistance.

$V = IR$ where:

V is **voltage**, measured in *volts* (V)

I is **current**, measured in *amperes* (A)

R is **resistance**, measured in *ohms* (Ω) (Ω is the Greek letter omega)

Question and Answer

An electric torch contains two 1·5 V batteries. The current in the bulb of the torch is 0·45 A. Calculate the resistance of the torch bulb.

$R = ?$

$I = 0{\cdot}45\ \text{A}$

$V = (1{\cdot}5 + 1{\cdot}5) = 3{\cdot}0\ \text{V}$

$V = IR$

$\Rightarrow 3{\cdot}0 = 0{\cdot}45 \times R$

$R = 6{\cdot}67$

$= 6{\cdot}7\ \Omega.$

Current and voltage in series circuits

Key Points

In a series circuit:

- the current is the ***same at all points*** in the circuit
- the ***sum of the potential differences*** across the components is ***equal to the voltage of the energy supply***.

(Learn these rules – and practise using them!)

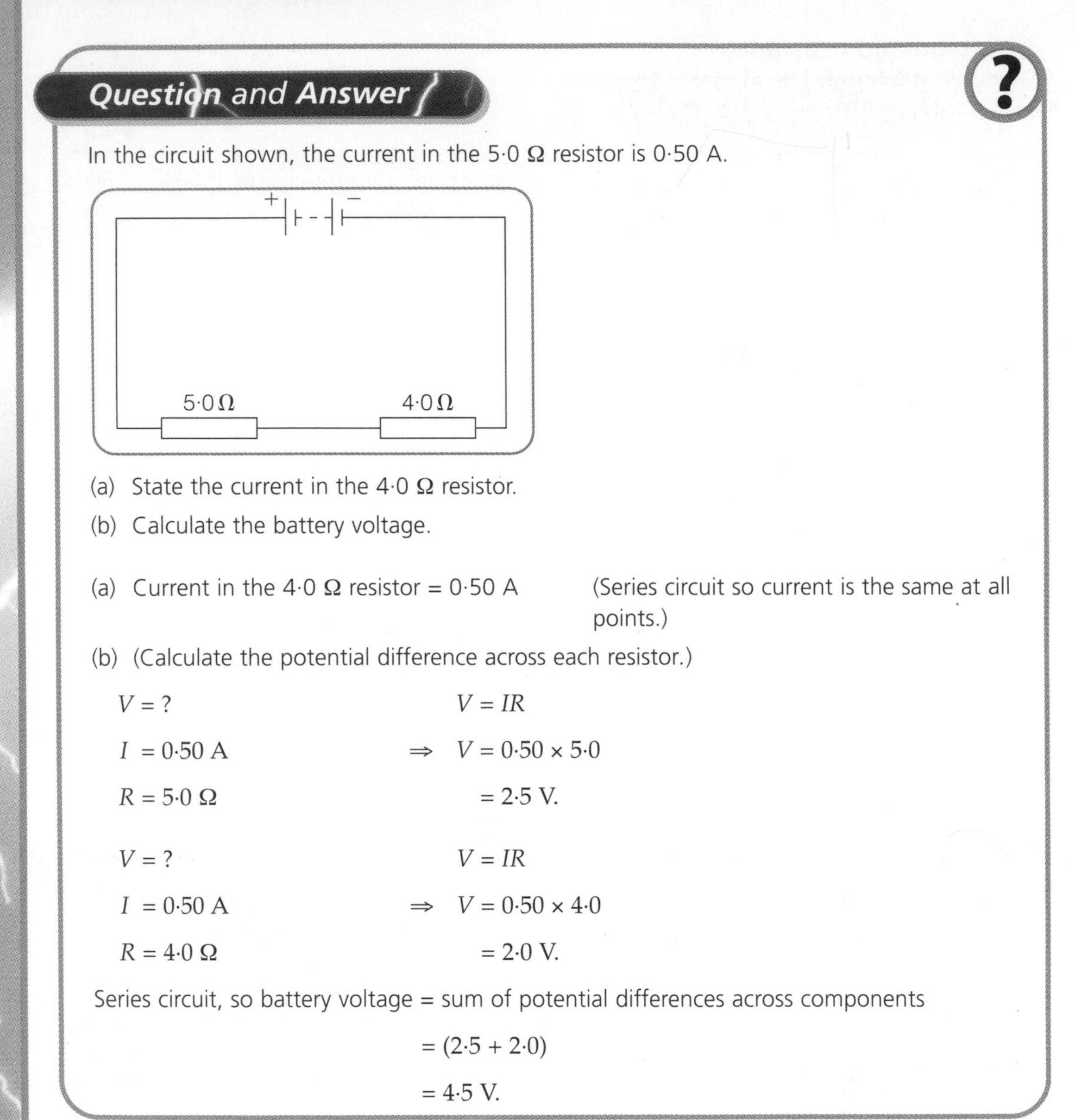

Question and Answer

In the circuit shown, the current in the 5·0 Ω resistor is 0·50 A.

(a) State the current in the 4·0 Ω resistor.

(b) Calculate the battery voltage.

(a) Current in the 4·0 Ω resistor = 0·50 A (Series circuit so current is the same at all points.)

(b) (Calculate the potential difference across each resistor.)

$V = ?$ $\quad\quad V = IR$

$I = 0{\cdot}50\ \text{A}$ $\quad\Rightarrow\quad V = 0{\cdot}50 \times 5{\cdot}0$

$R = 5{\cdot}0\ \Omega$ $\quad\quad = 2{\cdot}5\ \text{V}.$

$V = ?$ $\quad\quad V = IR$

$I = 0{\cdot}50\ \text{A}$ $\quad\Rightarrow\quad V = 0{\cdot}50 \times 4{\cdot}0$

$R = 4{\cdot}0\ \Omega$ $\quad\quad = 2{\cdot}0\ \text{V}.$

Series circuit, so battery voltage = sum of potential differences across components

$= (2{\cdot}5 + 2{\cdot}0)$

$= 4{\cdot}5\ \text{V}.$

Current and voltage in parallel circuits

Key Points

In a parallel circuit:

- the ***sum of the currents*** in the components is ***equal to the current in the energy supply***
- the potential difference across each component is ***equal to the voltage*** of the energy supply.

(Learn these rules – and practise using them!)

Question and Answer

In the circuit shown, the potential difference across the 3·0 Ω resistor is 9·0 V.

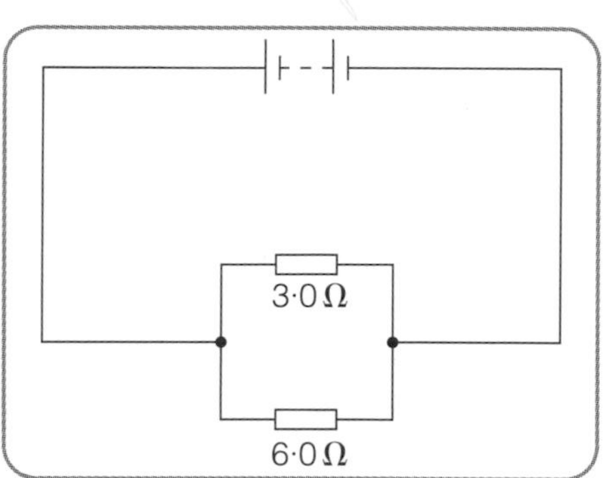

(a) State the battery voltage.

(b) Calculate the current in the 6·0 Ω resistor.

(a) Battery voltage = 9·0 V (Same as the p.d. across the 3·0 Ω resistor.)

(b) (p.d. across 6·0 Ω resistor = p.d. across 3·0 Ω resistor = battery voltage = 9·0 V)

$I = ?$ $\quad V = IR$

$V = 9{\cdot}0\ \text{V}$ $\quad \Rightarrow \quad 9{\cdot}0 = I \times 6{\cdot}0$

$R = 6{\cdot}0\ \Omega$ $\quad \Rightarrow \quad I = \dfrac{9{\cdot}0}{6{\cdot}0} = 1{\cdot}5\ \text{A}.$

In your Int2 exam you may be asked questions about circuits which have *both* series *and* parallel parts. There are questions like this in Exercises 12 and 13. You should use the solutions to these Exercises to learn how to tackle questions like these.

Exercises

Exercise 12 Current, voltage, and resistance

1 A student connects two resistors in series with a 12 V power supply, as shown in the circuit diagram. The current in the 18 Ω resistor is 0·25 A.

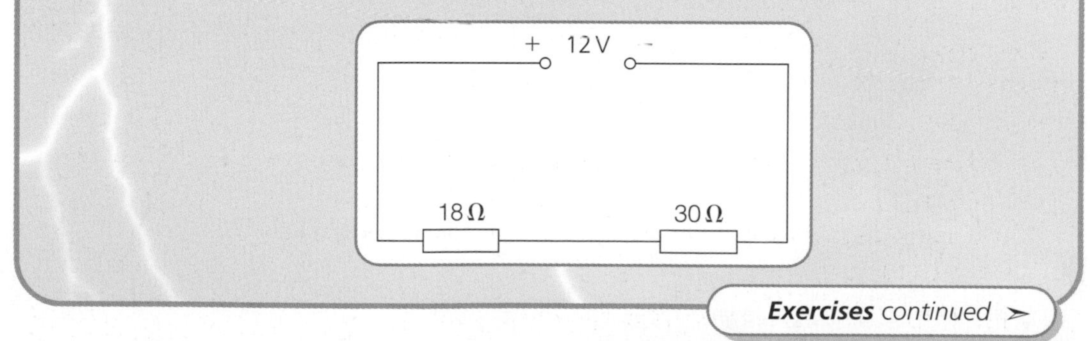

Exercises *continued* ➢

***Exercises** continued*

(a) Calculate the potential difference across the 30 Ω resistor.

(b) Calculate the total resistance of the circuit.

2 Two resistors are connected in parallel to a 6·0 V battery as shown.

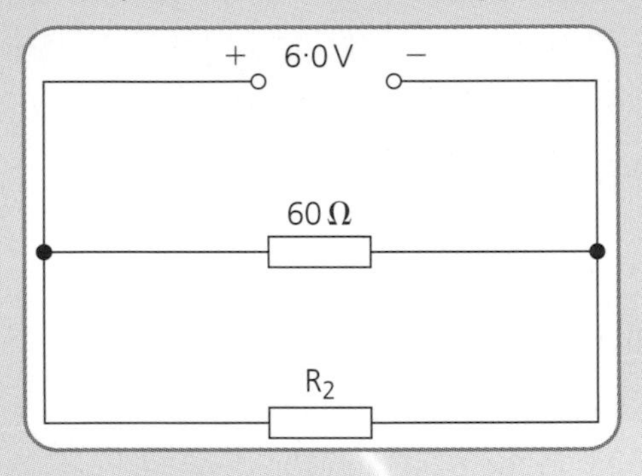

The current in the battery = 0·30 A.

(a) Calculate the current in the 60 Ω resistor.

(b) Calculate the resistance of resistor R_2.

3 Three resistors are connected to a 20 V supply as shown.

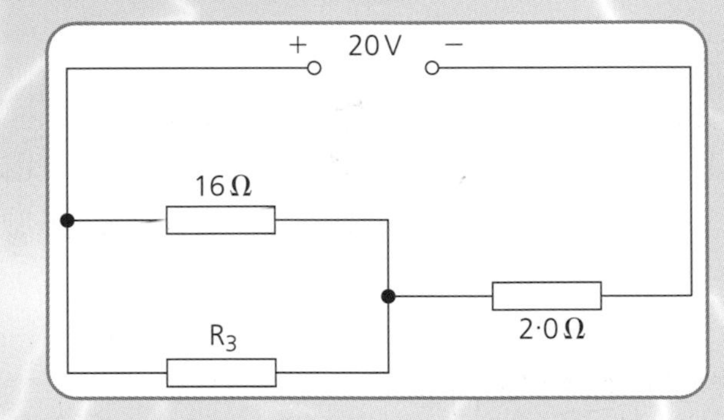

The current in the 2·0 Ω resistor is 2·0 A.

(a) State the current in the battery. Explain your answer.

(b) Calculate the p.d across the 16 Ω resistor.

(c) Calculate the resistance of resistor R_3.

3.3 Working with resistors

What You Should Know

For Intermediate 2 Physics you need to be able to:

- calculate the effective resistance of resistors connected in series and parallel
- describe potential dividers
- understand and use the term 'potential divider'
- carry out calculations on potential dividers.

Combining resistors

For two or more resistors connected in series, the total resistance is found using the relationship:

$$R_T = R_1 + R_2 + \dots .$$

(When another resistor is connected in series the total resistance goes up. There is a single path so the charge has to go through an extra resistor.)

For two or more resistors connected in parallel, the total resistance R_T is found using the relationship:

$$\frac{1}{R_T} = \frac{1}{R_1} + \frac{1}{R_2} + \dots .$$

(When another resistor is connected in parallel the total resistance goes down. There is an extra path for charge to follow so more charge can flow.)

When you use this relationship you ***must remember to invert after you work out the sum***.

Questions and Answers

1 Calculate the total resistance of the combination of resistors shown.

1·0 Ω 10 Ω 100 Ω

$R_1 = 1{\cdot}0\ \Omega$

$R_2 = 10\ \Omega$

$R_3 = 100\ \Omega$

$R_T = ?$

$R_T = R_1 + R_2 + R_3$

$\Rightarrow R_T = 1{\cdot}0 + 10 + 100$

$= 111\ \Omega.$

Questions and ***Answers*** *continued* ➢

Questions and Answers continued

2 Calculate the total resistance of the combination of resistors shown.

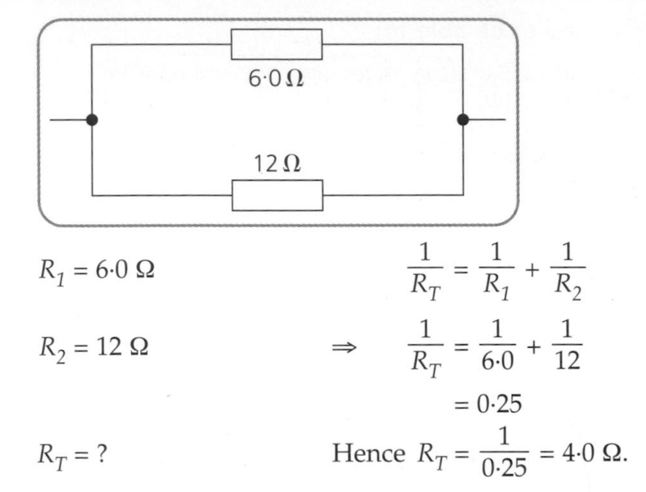

$R_1 = 6{\cdot}0\ \Omega$

$R_2 = 12\ \Omega$

$R_T = ?$

$$\frac{1}{R_T} = \frac{1}{R_1} + \frac{1}{R_2}$$

$$\Rightarrow \frac{1}{R_T} = \frac{1}{6{\cdot}0} + \frac{1}{12}$$

$$= 0{\cdot}25$$

$$\text{Hence } R_T = \frac{1}{0{\cdot}25} = 4{\cdot}0\ \Omega.$$

The potential divider

The potential divider consists of two (or more) resistors in series with a battery as shown in the diagram. By choosing the resistors carefully, we can control the voltage at point X.

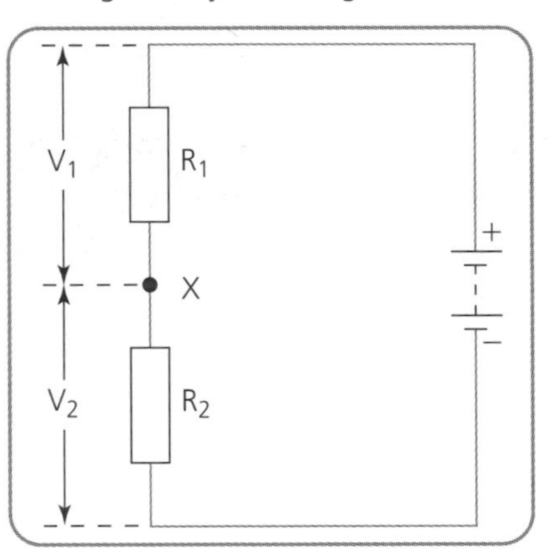

Because they are connected in series, the voltages across R_1 and R_2 are in the same ratio as the resistances. That is,

$$\frac{V_1}{V_2} = \frac{R_1}{R_2}.$$

The relationship for calculating the voltage across R_2 is $V_2 = \left(\frac{R_2}{R_1 + R_2}\right) V_s$, where:

V_2 is the **voltage** across R_2, measured in *volts* (V)

R_1 is the **resistance** of resistor R_1, measured in *ohms* (Ω)

R_2 is the **resistance** of resistor R_2, measured in *ohms* (Ω)

V_S is the **voltage** of the supply, measured in *volts* (V).

Potential dividers are used frequently in electronic circuits for controlling voltage.

Question and Answer

A potential divider is made by connecting two resistors in series with a 9·0 V battery as shown.

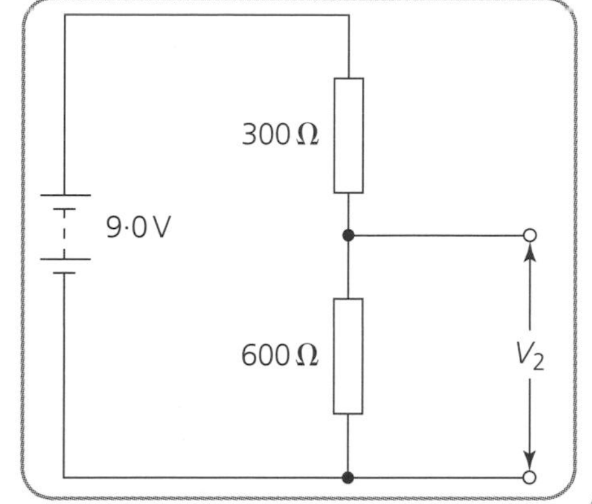

Calculate the voltage across the 600 Ω resistor.

$R_1 = 300\ \Omega$ $\quad V_2 = \left(\frac{R_2}{R_1 + R_2}\right)V_s$

$R_2 = 600\ \Omega$ $\quad \Rightarrow V_2 = \left(\frac{600}{300 + 600}\right) \times 9 \cdot 0$

$V_S = 9{\cdot}0\ \text{V}$ $\quad = 6{\cdot}0\ \text{V}.$

$V_2 = ?$

Exercises

Exercise 13 Working with resistors

1 A student is carrying out an experiment with ten 10 Ω resistors. The student connects all of the resistors firstly in series, and then in parallel.

(a) Calculate the total resistance of the ten resistors when they are connected in series.

(b) Calculate the total resistance of the ten resistors when they are connected in parallel.

2 A circuit is set up as shown.

(a) Calculate the total circuit resistance.

(b) Calculate the potential difference across the 8·0 Ω resistor.

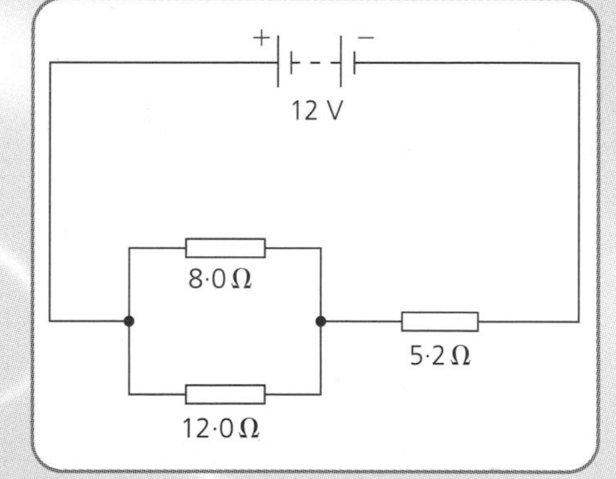

3 A circuit is set up as shown.

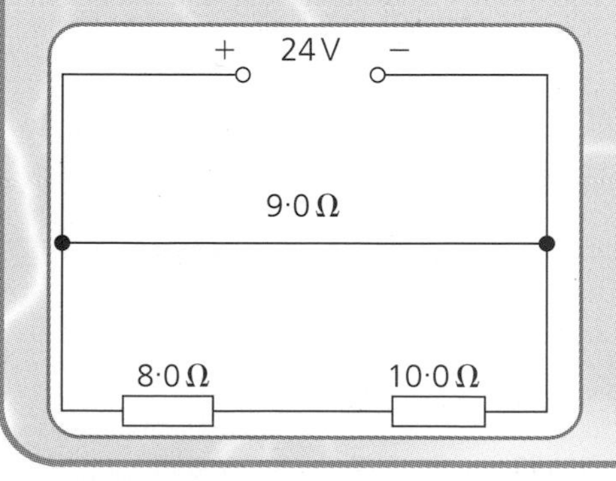

(a) Calculate the total circuit resistance.

(b) Calculate the potential difference across the 8·0 Ω resistor.

Exercises *continued* ➢

Exercises continued

4 A student sets up the following potential divider circuit.

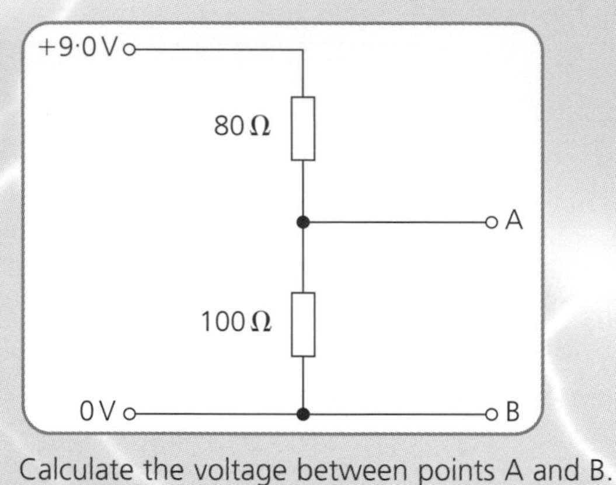

Calculate the voltage between points A and B.

3.4 Electrical energy

What You Should Know

For Intermediate 2 Physics you need to be able to:

- understand and use the equations $P = VI$, $P = I^2R$ and $P = V^2/R$
- carry out calculations on electrical power, electrical energy, and time
- identify energy changes in lamps and heaters
- understand and use the terms 'd.c.' and 'a.c.'
- state that the quoted voltage of an a.c. supply is less than its peak voltage
- state that a d.c. supply and an a.c. supply of the same voltage deliver the same power.

When there is a current in an electrical component, an energy change takes place. For example in a lamp, electrical energy is changed to heat and light. In the resistance wire of a heater, electrical energy is changed to heat. In some heaters light is also given out.

Calculating electrical power

The electrical energy changed to other forms each second is called the **power**. The SI unit of power is the *watt* and the symbol used by SQA for power is *P*.

In Int2 Physics, there are 3 equivalent relationships for calculating electrical power. Make sure you understand and know how to use all of them. They are:

$$P = VI, \qquad P = I^2R, \qquad P = \frac{V^2}{R}.$$

Question and Answer

A circuit is set up as shown.

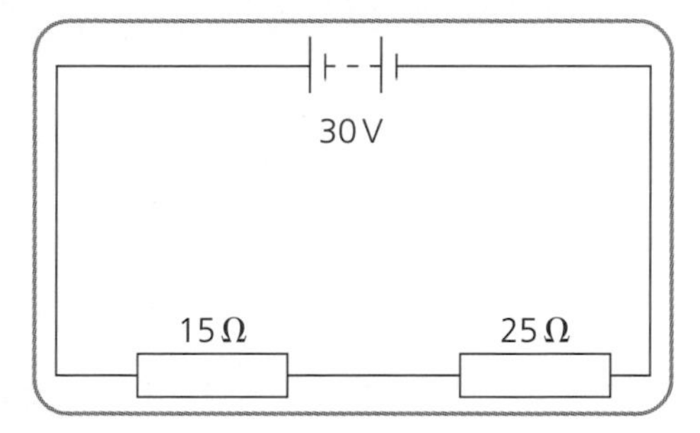

Calculate the energy transformed each second in the 25 Ω resistor.

(First calculate current in the circuit.) Total resistance of circuit = 15 Ω + 25 Ω = 40 Ω.

$R = 40\ \Omega$ $\quad V = IR$

$V = 30\ \text{V}$ $\quad \Rightarrow 30 = I \times 40$

$I = ?$ $\quad \Rightarrow I = 0{\cdot}75\ \text{A}.$

$I = 0{\cdot}75\ \text{A}$ $\quad$ Now using $P = I^2R$

$R = 25\ \Omega$ $\quad P = (0{\cdot}75)^2 \times 25$

$P = ?$ $\quad = 14{\cdot}06$

$= 14\ \text{W}.$

d.c. and a.c. energy supplies

The abbreviation 'd.c.' which is used for electrical supplies stands for ***direct current***. A d.c. supply causes charges to move always in the same direction around a circuit. That is, the current is always in the same direction. For example a battery is a d.c. supply.

The abbreviation 'a.c.' which is also used for electrical supplies stands for ***alternating current***. An a.c. supply causes charges to change direction when they are moving in a circuit. That is, the current changes direction frequently. For example mains electricity is an a.c. supply.

In an a.c. supply, the voltage of the supply goes through a regular cycle of positive and negative voltages. In the mains supply, there are fifty complete cycles of positive and negative voltage each second.

We say the frequency of the mains supply is 50 Hz.

The quoted value of the voltage of an a.c. supply is less than the peak value of its voltage.

When we draw a graph of mains a.c. voltage against time, it has the following appearance.

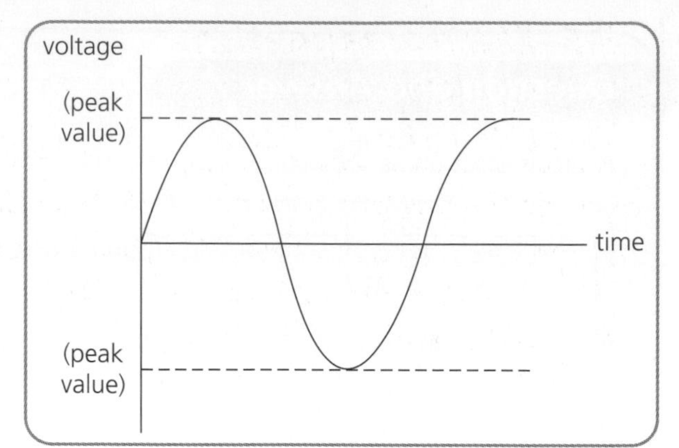

An a.c. supply and a d.c. supply of the same quoted voltage will both supply the same power to a resistor.

Question and Answer

A student has a d.c. supply with a fixed output of 12 V. The student also has an a.c. supply of variable output. What setting of the variable a.c. supply will deliver the same power as the d.c. supply?

The variable a.c. supply should be set to 12 V.

Exercises

Exercise 14 Electrical energy

1 In an electric fire where does the energy change take place?

2 State the energy change in a filament lamp.

3 (a) Starting from $P = IV$, show that $P = I^2R$.

(b) Starting from $P = IV$, show that $P = \frac{V^2}{R}$.

4 A label on an electric kettle contains the following data, 'Power 3·0 kW Voltage 230 V'.

(a) Calculate the current in the kettle when it is on.

(b) How much electrical energy does the kettle use in 5 minutes?

5 A mains operated electric toaster has four identical elements connected in parallel as shown in the diagram.

The power rating of the toaster is 1·84 kW.

Calculate the resistance of each element.

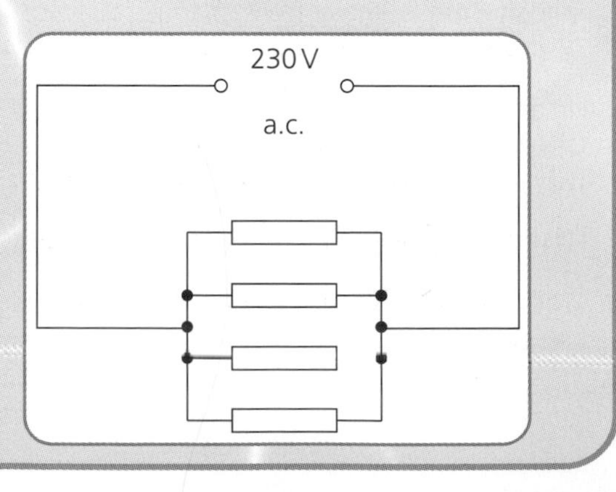

3.5 Electromagnetism

What You Should Know

For Intermediate 2 Physics you need to be able to:

- state that there is magnetic field around a wire carrying electric current
- identify when a voltage is induced in a conductor
- state the factors which affect the induced voltage
- state that transformers are used to change the sizes of alternating voltages
- carry out calculations on transformers.

Induction

When there is an electric current in a wire, there is a magnetic field around the wire. The motion of the charges causes this magnetic field.

When a wire is moved ***across the lines of a magnetic field*** a voltage is set up in the wire. The motion of the wire across the magnetic field causes this voltage.

We say the voltage is ***induced*** in the wire.

When the wire is moved **along the lines of a magnetic field** no voltage is induced.

When the ***direction*** of movement is ***reversed***, the direction of the ***induced voltage*** is ***reversed***.

When the wire is wound into a coil the induced voltage is greater.

The size of the induced voltage is also greater when

- the magnetic field is ***stronger***
- there are ***more turns*** in the coil
- the coil is moved ***faster*** across the lines of the magnetic field.

In your Int2 exam you could be asked about these factors, so learn them!

Transformers

Transformers are used to change the size of a.c. voltages.

A transformer consists of two coils of wire linked by an iron core. (The iron core improves the magnetic link between the two coils.)

When there is a changing current in one (the **primary**) coil of a transformer, a changing voltage is induced in the other (the **secondary**) coil.

The size of the induced voltage depends on the numbers of turns in the coils.

The frequency of the induced voltage is the same as the frequency of the voltage in the primary coil.

The relationship between the numbers of turns in the coils and the voltages is

$$\frac{V_s}{V_p} = \frac{n_s}{n_p} \quad \text{where:}$$

V_s is the **voltage induced** in the secondary coil, measured in *volts* (V)

V_p is the **voltage** across the primary coil, measured in *volts* (V)

n_s is the **number of turns** in the secondary coil (it's just a number so no unit)

n_p is the **number of turns** in the primary coil.

For a transformer which is 100% efficient, the *power in the secondary circuit* is equal to the *power in the primary circuit*. Hence $V_s I_s = V_p I_p$,

$$\Rightarrow \frac{V_s}{V_p} = \frac{I_p}{I_s}.$$

You do not need to be able to derive this relationship, but you do need to know how to use it.

Combining it with the original relationship, we have

$$\frac{V_s}{V_p} = \frac{I_p}{I_s} = \frac{n_s}{n_p}.$$

Question and Answer

A 230 V a.c. supply is connected to the primary coil of a transformer as shown.

(a) Calculate the voltage induced in the secondary coil.

(b) Calculate the current in the secondary circuit.

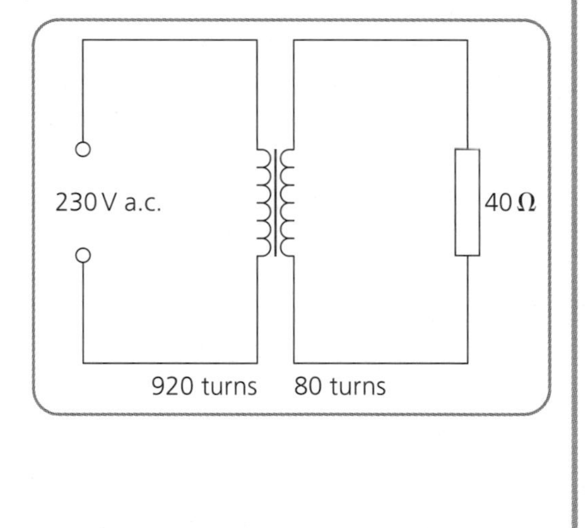

(a) $V_p = 230$ V $\qquad \dfrac{V_s}{V_p} = \dfrac{n_s}{n_p}$

$n_p = 920 \qquad \Rightarrow \dfrac{V_s}{230} = \dfrac{80}{920}$

$n_s = 80 \qquad \Rightarrow V_s = 20$ V.

$V_s = ?$

(b) $V_s = 20$ V $\qquad V = IR$

$R = 40\ \Omega \qquad \Rightarrow 20 = I_s \times 40$

$I_s = ? \qquad \Rightarrow I_s = 0{\cdot}5$ A.

Exercises

Exercise 15 Electromagnetism

1 A 25 V d.c. supply is connected to the primary coil of a transformer. State the voltage induced in the secondary coil. Explain your answer.

2 A 20 V a.c. supply is connected to the primary coil of a transformer as shown.

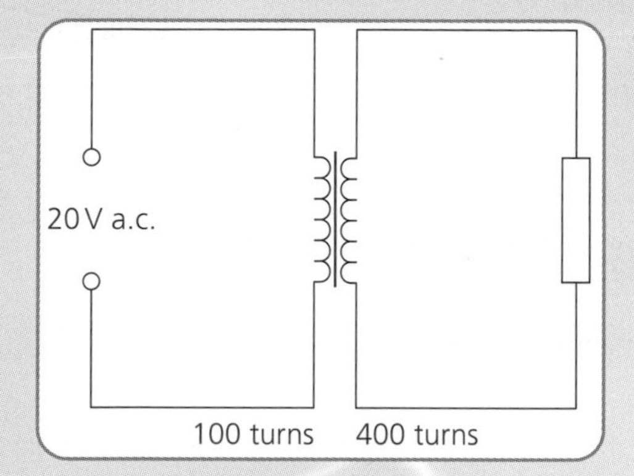

(a) Calculate the voltage induced in the secondary coil.

(b) The current in the secondary circuit is 0·40 A. Calculate the resistance of the resistor in the secondary circuit.

3 A mains transformer is used in the power supply to a model train set. The maximum power used by the model train is 65 W. Calculate the current in the primary coil of the transformer when the model train is moving at its maximum speed. (Assume that the mains voltage is 230 V a.c.)

4 An a.c. supply is connected to the primary coil of a transformer as shown.

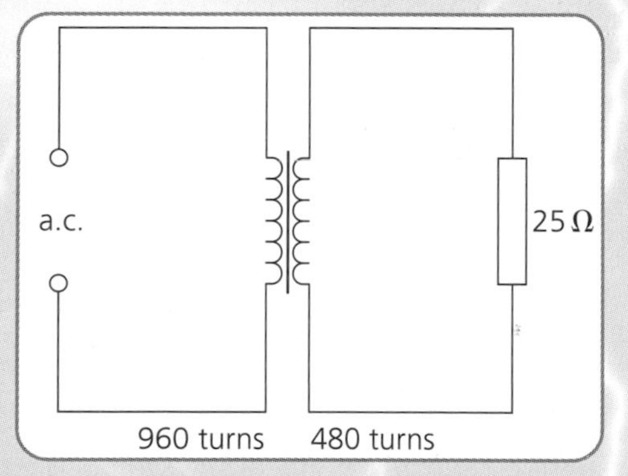

The current in the secondary circuit is 2·0 A.

(a) Calculate the power used in the secondary circuit.

(b) Calculate the current in the primary circuit.

(c) Calculate the voltage of the supply.

3.6 Electronic components

What You Should Know

For Intermediate 2 Physics you need to be able to:

- give examples of output devices, input devices, and their energy changes
- draw and identify the symbol for an LED and explain how to use an LED in a circuit
- carry out calculations for thermistors and LDRs in circuits.

Output devices

An *output device* changes electrical energy to one or more other forms of energy.

The following table includes details on common output devices.

Output device	Symbol used by SQA	Energy change
LED (light emitting diode)		electrical energy → light
buzzer		electrical energy → sound
motor	M	electrical energy → kinetic energy
loudspeaker		electrical energy → sound

You need to be able to identify and draw the symbol for an LED (light emitting diode).

You also need to know about the energy changes in the various output devices.

The following circuit shows an LED connected correctly in a circuit.

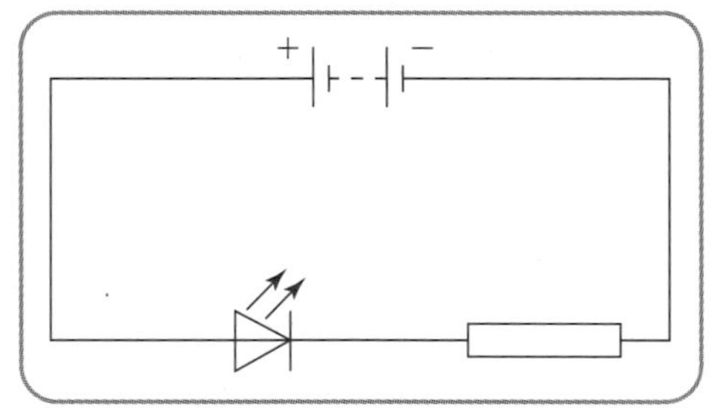

Study this diagram carefully! Note which side of the symbol is connected to the positive side of the battery, and which side is connected to the negative side.

An LED only lights if it is connected the right way round.

Many Int2 students lose marks by drawing LED symbols the wrong way round. Make sure *you* get it right!

The resistor protects the LED by making sure that the current is not too big.

Question and Answer

An LED, connected to a 5·0 V battery as shown in the previous circuit diagram is operating correctly. The potential difference across the LED is 1·6 V when it is on its operating current of 20 mA. Calculate the resistance of the resistor.

$I = 20\text{ mA} = 0{\cdot}020\text{ A}$ $\qquad V = IR$

$V = 5{\cdot}0 - 1{\cdot}6 = 3{\cdot}4\text{ V}$ $\qquad \Rightarrow \quad 3{\cdot}4 = 0{\cdot}020 \times R$

$R = ?$ $\qquad \Rightarrow \quad R = 170\ \Omega.$

Input devices

An *input device* changes another form of energy to electrical energy.

The following table includes details on common input devices.

Input device	Symbol used by SQA	Energy change
microphone		sound → electrical energy
thermocouple	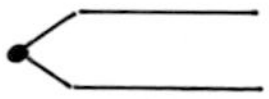	heat → electrical energy
solar cell		heat + light → electrical energy
thermistor		heat → electrical energy
LDR (light dependent resistor)		light → electrical energy

The resistance of most common thermistors *decreases* as temperature *rises*.

The resistance of an LDR (light dependent resistor) *decreases* when there is *more light* shining on the LDR.

You need to be able to solve problems on circuits which include thermistors and LDRs.

Question and Answer

Part of the circuit diagram of an electronic system is shown in the diagram.

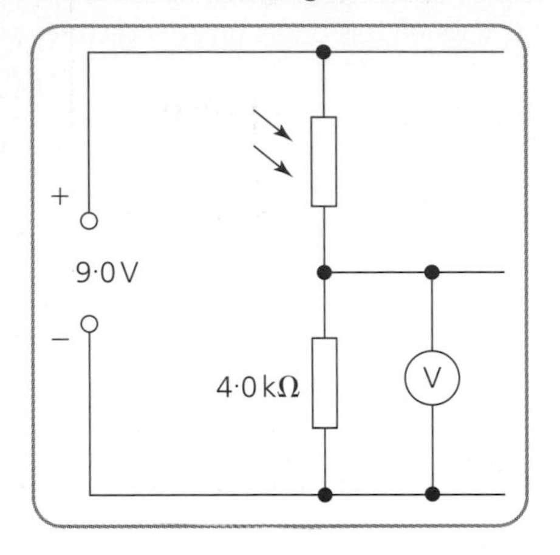

The reading on the voltmeter is 2·0 V

Calculate the resistance of the LDR.

$V_2 = 2{\cdot}0\ \text{V}$

$V_S = 9{\cdot}0\ \text{V}$

$R_2 = 4{\cdot}0\ \text{k}\Omega = 4000\ \Omega$

$R_1 = ?$

$$V_2 = \left(\frac{R_2}{R_1 + R_2}\right)V_S$$

$$\Rightarrow 2{\cdot}0 = \left(\frac{4000}{R_1 + 4000}\right) \times 9{\cdot}0$$

$$\Rightarrow 2{\cdot}0 \times (R_1 + 4000) = 4000 \times 9{\cdot}0$$

$$\Rightarrow 2R_1 + 8000 = 36\,000$$

$$\Rightarrow R_1 = 14\,000\ \Omega.$$

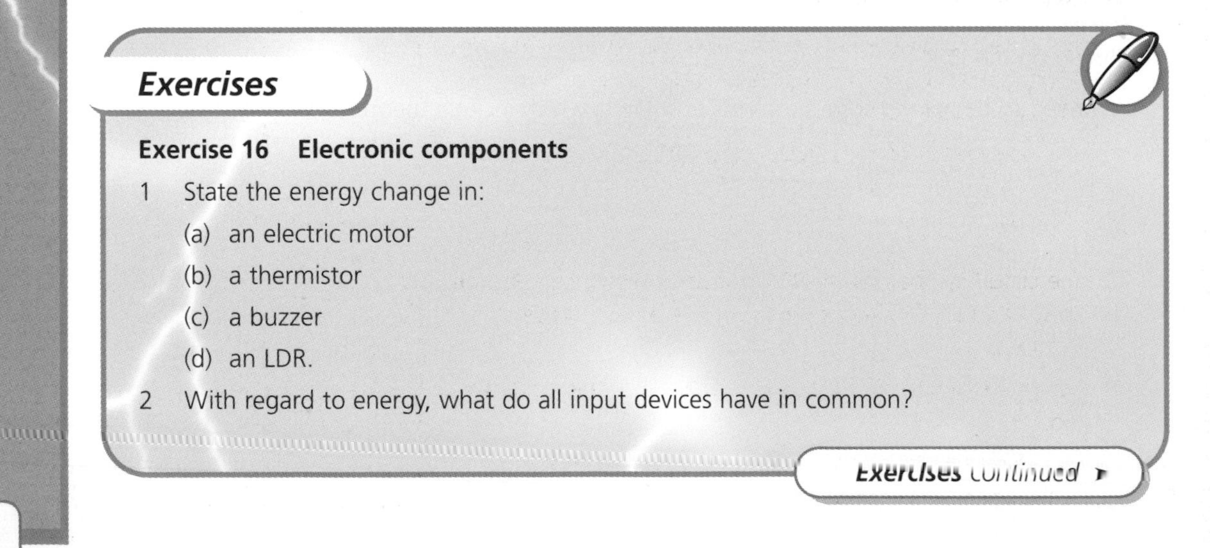

Exercises

Exercise 16 Electronic components

1 State the energy change in:
   (a) an electric motor
   (b) a thermistor
   (c) a buzzer
   (d) an LDR.

2 With regard to energy, what do all input devices have in common?

Exercises continued ➤

Exercises continued

3 Part of the circuit diagram of an electronic system is as shown.

The resistance of component X is 2·5 kΩ.

(a) Name component X.

(b) Calculate the p.d. between points A and B.

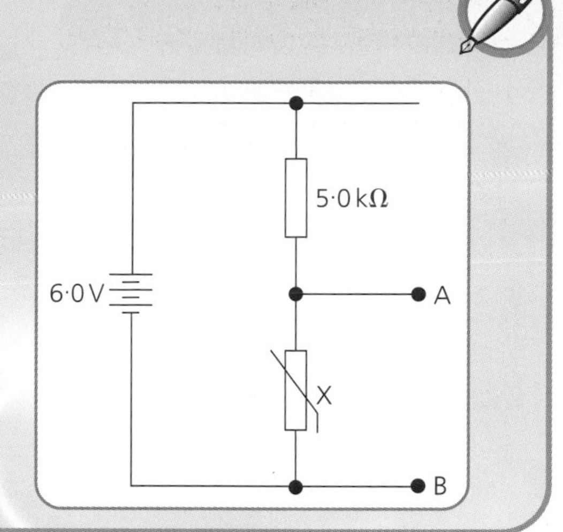

3.7 Electronic switches and amplifiers

What You Should Know

For Intermediate 2 Physics you need to be able to:

- draw and identify the symbols for electronic switching devices
- explain the operation of a simple transistor switch
- identify devices which use amplifiers
- carry out calculations on amplifiers.

Electronic switching devices

1. The circuit symbol for an n-channel enhancement MOSFET which can be used as a very fast electronic switch is shown in the diagram.

   You need to be able to recognise and draw this symbol.

2. The circuit symbol for an NPN transistor which can also be used as a very fast electronic switch is shown in the diagram.

   You need to be able to recognise and draw this symbol too.

   You also need to be able to explain how a transistor circuit operates as a switch.

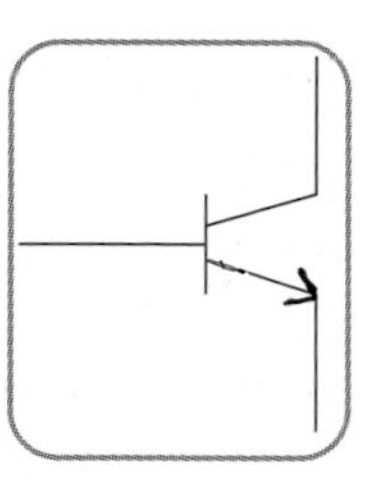

Question and Answer

The circuit shown is used to switch on the buzzer when the temperature rises above -3°C.

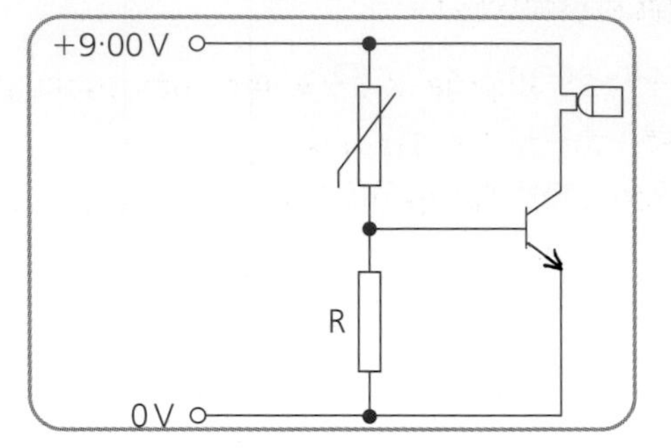

(a) Explain how this circuit operates to switch on the buzzer.

(b) Suggest **one** use for this circuit.

(a) (When the temperature is below -3°C the transistor is off.)

As the temperature increases the resistance of the thermistor falls.

The voltage across resistor R therefore increases.

When the voltage across resistor R rises above 0·7 V the transistor switches on and there is then a current in the buzzer.

(b) This circuit could sound an alarm when the temperature inside a freezer becomes too high.

Amplifiers

An *amplifier* is used to increase the energy of an electronic signal. The amplifier increases the *amplitude* of the electronic signal. The frequency of the signal does not change.

Amplifiers are commonly used in radios, televisions, record players, electronic music players … and so on.

In all of these devices the amplifier is used to increase the energy of an audio signal before it is passed to loudspeakers or earphones.

The *gain* of an amplifier is a measure of the increase in the amplitude of the signal.

There are two relationships for calculating the gain of an amplifier. These are:

$$V_{\text{gain}} = \frac{V_o}{V_i} \quad \text{and} \quad P_{\text{gain}} = \frac{P_o}{P_i}\text{, where:}$$

V_{gain} is **voltage gain**, which is a number (it has *no unit*)

V_o is **output voltage**, measured in *volts* (V)

V_i is **input voltage**, measured in *volts* (V)

and P_{gain} is **power gain**, which is a number (it has *no unit*)

P_o is **output power**, measured in *watts* (W)

P_i is **input power**, measured in *watts* (W).

In your Int2 exam you might be asked to use either of these relationships.

Question and Answer

The input voltage to an audio amplifier is 0·6 V The output voltage is 3·0 V.

(a) Calculate the voltage gain of this amplifier.

(b) An audio amplifier does not change the frequency of the input signal. What observation confirms this?

(a) $V_o = 3{\cdot}0$ V $\qquad V_{gain} = \dfrac{V_o}{V_i}$

$V_i = 0{\cdot}6$ V $\qquad \Rightarrow V_{gain} = \dfrac{3{\cdot}0}{0{\cdot}6} = 5$

$V_{gain} = ?$ $\qquad = 5$ *(Voltage gain does not have a unit.)*

(b) After amplification, the pitch of a note in music, or the sounds of peoples voices do not sound higher or lower than they should.

Exercises

Exercise 17 Electronic switches and amplifiers

1 The circuit shown is used to switch off a relay when the intensity of light incident on the LDR rises to a certain level. The relay then switches off a lamp.

During the hours of darkness the resistance of the LDR is 5 kΩ and the voltage across the LDR is 6·0 V.

(a) Name component X.

(b) Calculate the resistance of resistor R.

(c) Describe how the circuit operates to switch off the relay.

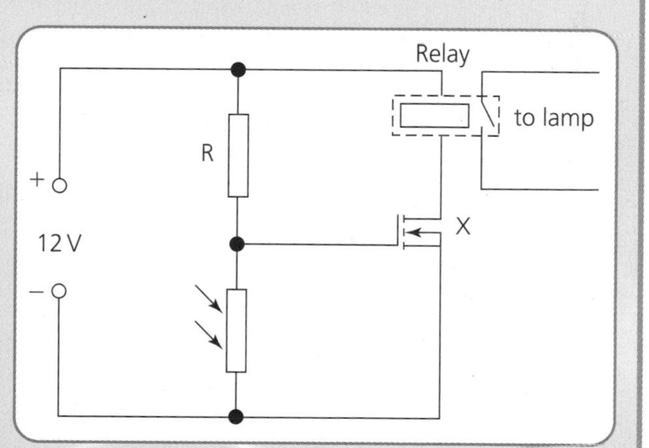

2 A signal with a frequency of 250 Hz is applied to the input of an audio amplifier with a power gain of 10. The input power to the amplifier is 1·2 W.

(a) State the frequency of the output signal of the amplifier. Explain your answer.

(b) Calculate the output power of the amplifier.

Chapter 4

WAVES AND OPTICS

4.1 Waves

What You Should Know

For Intermediate 2 Physics you need to be able to:

- understand and use the terms 'wave', 'wavelength', 'speed', 'amplitude', 'frequency', and 'period'
- state that a wave carries energy
- describe the difference between longitudinal and transverse waves
- state the speed of radio, television, and light waves
- carry out calculations using the equations $d = vt$ and $v = f\lambda$ for different kinds of waves
- describe how to measure the speed of sound
- state the order of the electromagnetic spectrum.

Basic concepts

Make sure you understand all of the following concepts and terms. Any of these could come up in your exam.

Key Words and *Definitions*

A *wave* is a regular vibration which carries energy from one place to another. For example, ripples on the surface of a pond carry energy which causes objects floating on the surface of the pond to move up and down.

The following diagram shows a **transverse** wave.

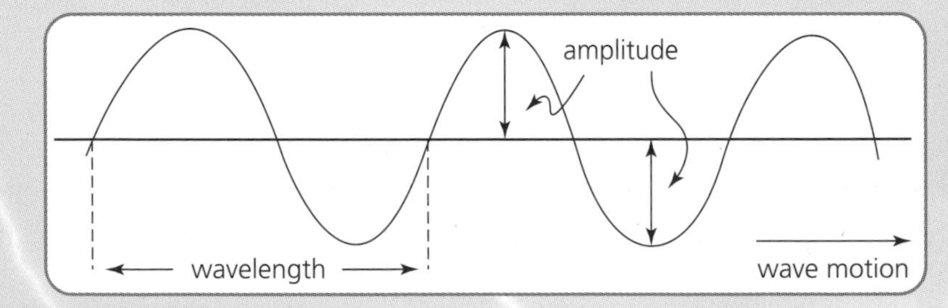

In a *transverse wave* the vibration is at *right angles* to the direction in which the wave is moving. Waves on water, and light waves are transverse waves.

Key Words and **Definitions** continued ➢

Key Words and *Definitions* continued

In *longitudinal waves* the vibration is *parallel* to the direction in which the wave is moving. Sound and ultrasound are longitudinal waves.

The *amplitude* of a wave is the maximum displacement of the vibration. Increasing the amplitude of a wave increases the energy carried by the wave.

The **wavelength** of a wave is the length of one complete wave. The SI unit of wavelength is the *metre* and the symbol used by SQA for wavelength is λ.

The **frequency** of a wave is the number of waves made each second. It is also equal to the number of waves passing a point each second. The SI unit of frequency is the *Hertz* and the symbol used by SQA for frequency is f.

The *period* of a wave is the time to make one complete wave.

Wave speed

The speed of a wave is how fast the wave moves.

For example, light waves, radio waves, and television waves all move at a speed of 3×10^8 m/s (300 million m/s) through air.

The speed of sound in air is 340 m/s.

In Int2 Physics there are two relationships which include the speed of waves. These are:

$$d = vt \quad \text{and} \quad v = f\lambda .$$

You have to be able to use *both* of these relationships.

Question and *Answer*

The Earth is approximately $1{\cdot}5 \times 10^{11}$ m from the Sun. Calculate the time, in minutes, for light to travel from the Sun to Earth.

(Speed of light in vacuum = 3×10^8 m/s.)

$d = 1{\cdot}5 \times 10^{11}$ m

$v = 3{\cdot}0 \times 10^8$ m/s

$t = ?$

Using $d = vt$,

$1{\cdot}5 \times 10^{11} = 3{\cdot}0 \times 10^8 \times t$

$\Rightarrow t = 500 \text{ s} = 8{\cdot}3$ minutes.

You must also be able to describe a method for measuring the *speed of sound* using the equation $d = vt$.

Question and Answer

Describe a method for measuring the speed of sound.

1 Set up the apparatus as shown.

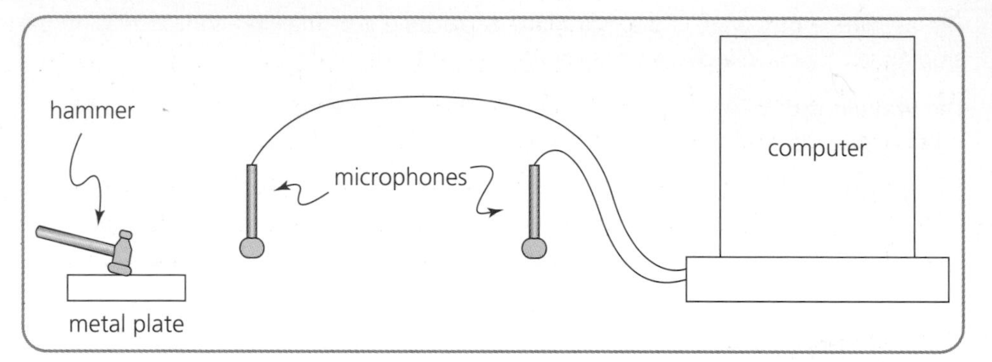

The microphones are placed a distance apart. This distance is measured at least five times, and an average distance is calculated.

2 The hammer strikes the metal plate to make a sound.

The computer measures the time interval between the sound arriving at each microphone. This time measurement is taken at least five times, and an average time is calculated.

3 The speed of sound is then calculated by dividing the measured distance by the measured time.

The Electromagnetic Spectrum

The electromagnetic spectrum is a family of waves.

All electromagnetic waves move at the same speed in air. The waves have different frequencies and wavelengths.

The following diagram shows the electromagnetic spectrum in order of increasing wavelength. Gamma rays have the shortest wavelengths, and radio waves the longest wavelengths.

gamma rays	X-rays	ultraviolet rays	visible light	infrared rays	microwaves	TV waves	radio waves

For Int2 Physics you need to be able to list these waves in the order of their wavelengths, so study the diagram carefully.

Exercises

Exercise 18 Waves

1 (a) Define the frequency of a longitudinal wave.

(b) Give an example of a longitudinal wave.

2 Ripples on the surface of a canal take 35 s to travel a distance of 7·0 m. A girl on the canal bank counts 20 ripples passing her in a time of 10 s.

Calculate the wavelength of the ripples.

3 A radio wave has a wavelength of 250 m. Calculate the frequency of this wave.

4 Name two electromagnetic waves with wavelengths longer than visible light.

4.2 Reflection

What You Should Know

For Intermediate 2 Physics you need to be able to:

- understand and use the terms 'angle of incidence', 'angle of reflection', 'normal'
- state that the path followed by a light ray is the same for both directions of travel
- explain the effect of curved reflectors on received and transmitted signals
- describe an application of curved reflectors in telecommunication
- explain the terms 'total internal reflection' and 'critical angle'
- describe how an optic fibre transmits light rays.

Plane mirrors

A ray of light incident on a flat (plane) mirror is reflected as shown in the diagram.

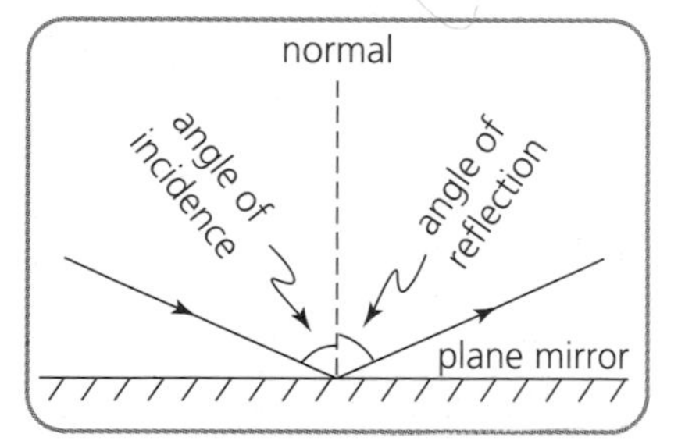

The *normal* is a line perpendicular to the mirror.

The *angle of incidence* is the angle between the incident ray and the normal. The *angle of reflection* is the angle between the reflected ray and the normal. Remember:

angle of incidence = angle of reflection.

When a ray of light is reflected back along its path, it follows exactly the same path back to its source. This is called the *principle of reversibility* of ray paths. You do not need to remember this name but you do need to be able to state the principle.

Curved reflectors

When parallel rays are incident on a curved reflector, the rays are reflected such that they all pass through a single point.

This point is called the **focus**.

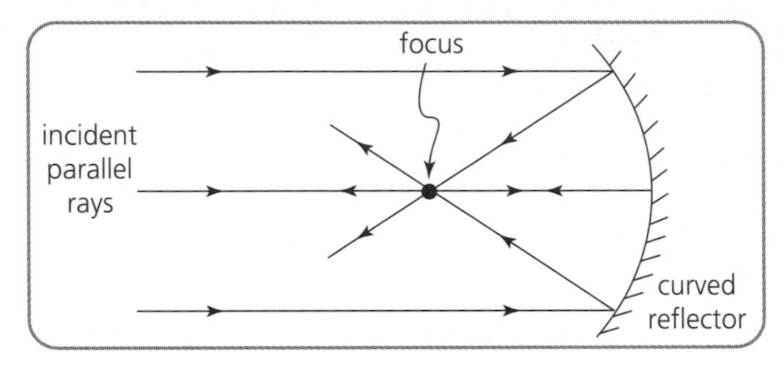

All of the wave energy which hits the curved reflector passes through the focus. The wave signal is strongest at the focus.

A curved reflector like this is useful for receiving weak signals.

Question and Answer

A lamp, placed at the focus of a curved reflector, sends out light rays in all directions.

(a) By referring to the previous diagram explain the effect of the curved reflector on the light rays which strike the reflector.

(b) Hence draw a diagram showing the paths of the reflected rays.

(a) The principle of reversibility of ray paths means that rays of light going in the opposite directions to those in the previous diagram will follow exactly the same paths.

Rays which are initially diverging from the focus are reflected into a parallel beam.

(b)

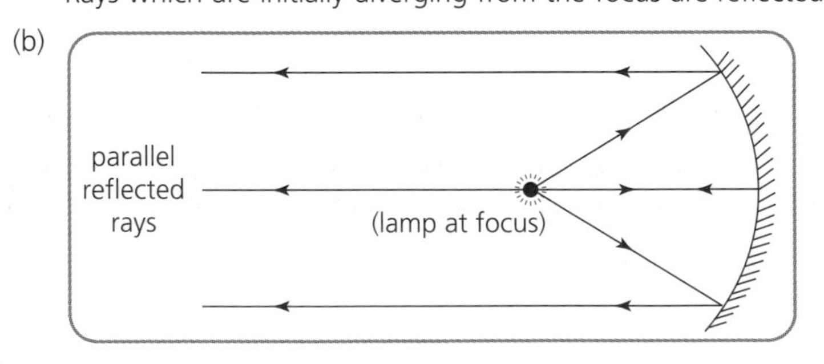

Total internal reflection

When light is incident on the internal surface of a transparent material like glass, some of the light is reflected. The following diagrams show the changes in reflection as the angle of incidence in the glass is increased.

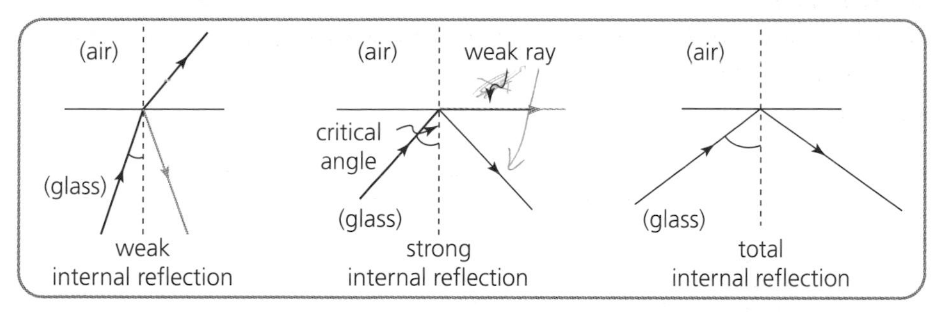

When the angle of incidence is small, a little light energy is reflected. Most of the light passes into the air.

As the angle of incidence is increased, more of the light energy is reflected.

The middle diagram shows the *critical angle of incidence*. (At this angle, very little light passes into the air.)

When the angle of incidence is bigger than the critical angle, all of the light energy is reflected inside the material. This is called **total internal reflection**.

It is called **total** because *all* of the energy is reflected, and **internal** because the energy stays *inside* the material.

This principle is used in optical fibres

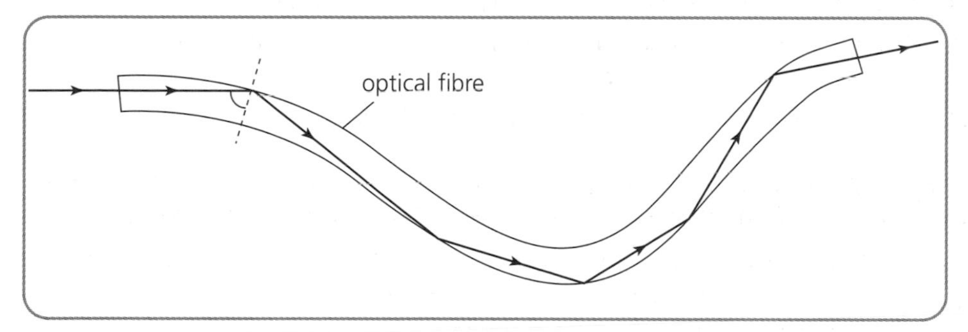

A ray enters the fibre at one end. The ray is repeatedly reflected inside the fibre.

For each reflection, the angle of incidence between the ray and the normal to the side of the fibre is greater than the critical angle, so the ray stays inside the fibre until it reaches the other end.

Exercises

Exercise 19 Reflection

1 Describe a use for curved reflectors in telecommunications.

2 A student directs a ray of red light at a triangular prism as shown in the diagram.

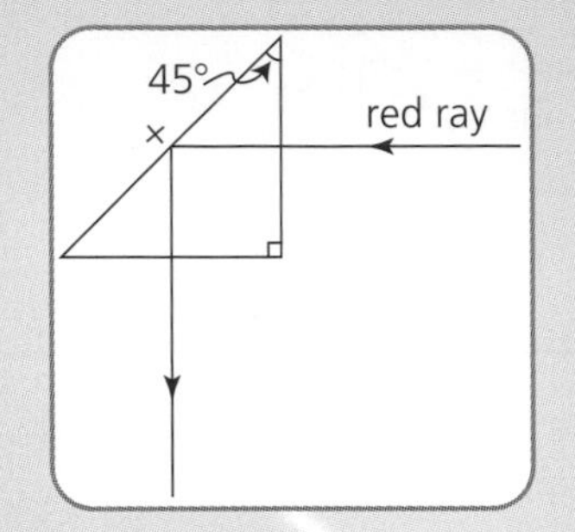

(a) Name the effect which occurs at point X.

(b) The student replaces the prism with second prism, as shown.

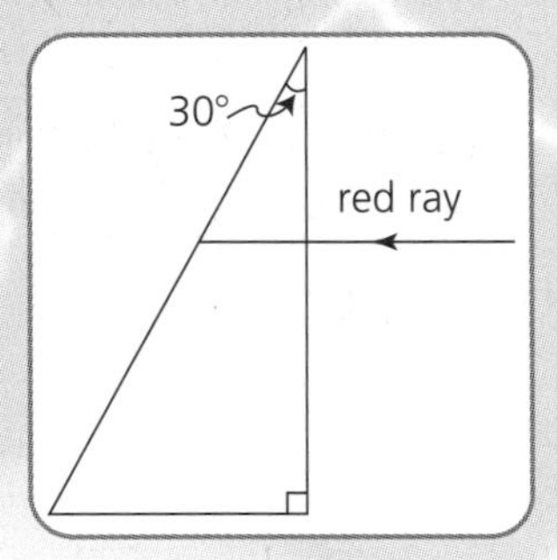

Does the same effect occur? Explain your answer.

4.3 Refraction

What You Should Know

For Intermediate 2 Physics you need to be able to:

- understand and use the term 'angle of refraction'
- draw diagrams to show light passing from air to glass and glass to air
- describe converging and diverging lenses and their effects on parallel rays
- draw diagrams to show how a converging lens forms an image
- carry out calculations on the power of a lens
- describe long and short sight and explain how they are corrected using lenses.

Refraction occurs when a wave moves from one material to another.

A ray of light passing through a rectangular block of glass is refracted as shown in the diagram.

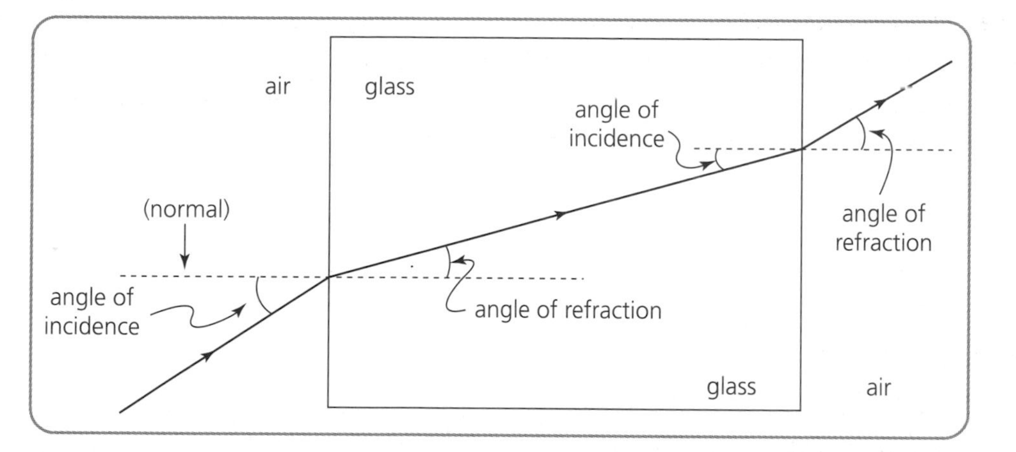

Key Points

The *angle of refraction* is the angle between the refracted ray and the normal.

For a ray moving from air to glass *angle of incidence* > *angle of refraction*.

For a ray moving from glass to air *angle of incidence* < *angle of refraction*.

Converging lenses

A *converging lens* is 'fatter' in the middle and 'thinner' at the edges.

When parallel rays pass through a converging lens they are made to converge. That is they come together.

After leaving the lens, all of the rays pass through a single point called the *focus*.

The distance from the lens to the focus is called the *focal length*, *f*, of the lens.

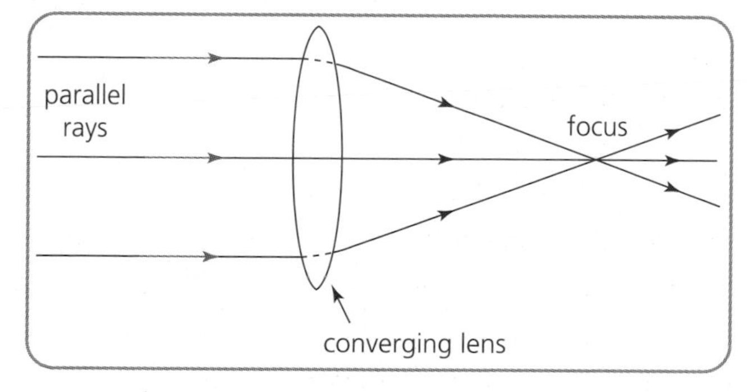

For Int2 Physics you need to be able to draw diagrams to show how a converging lens forms the image of an object at various distances defined in terms of focal length.

Question and Answer

Draw a diagram to show how a converging lens forms the image of an object between one and two focal lengths from the lens.

The completed diagram looks like this.

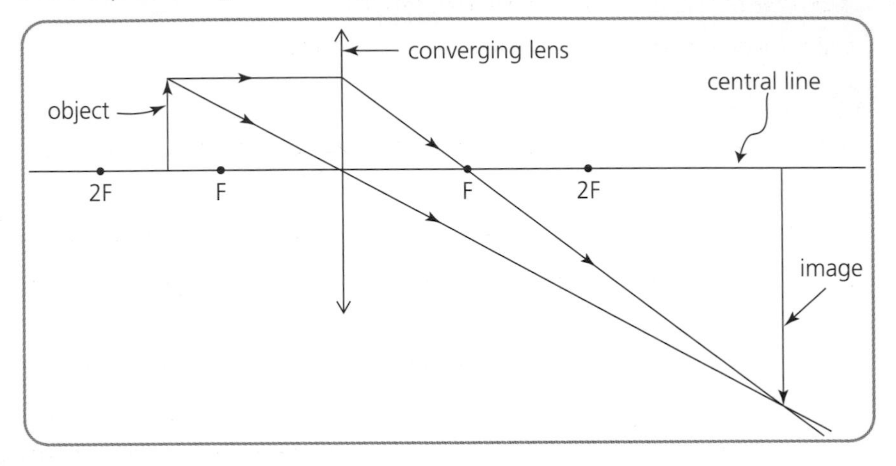

It has been produced using the following steps:

1. Draw a converging lens in the middle of the page. (A straight line will do – you can put little caps on each end to show that the edges of the lens are thinner than the middle.)
2. Draw a straight line through the centre of the lens – this is the central line, or axis.
3. Mark four points on the central line at **one** and **two** focal lengths on **both sides** of the lens.
4. Now, draw a vertical arrow on one side of the central line – this represents the object.
5. Draw a **ray** from the top of the arrow through the centre of the lens – put arrows on it to show that it is a ray. This ray, is undeviated.
6. Draw a **ray** parallel to the central line, from the top of the arrow to the lens. This ray is 'turned' so that it goes through the focal point one focal length from the lens on the right hand side. Extend this ray until it crosses the ray drawn through the centre of the lens.
7. The image of the top of the object is formed where the two rays cross.

Diverging lenses

A diverging lens is 'fatter' at the edges and 'thinner' in the middle.

When parallel rays pass through a diverging lens they are made to diverge. That is they get further apart.

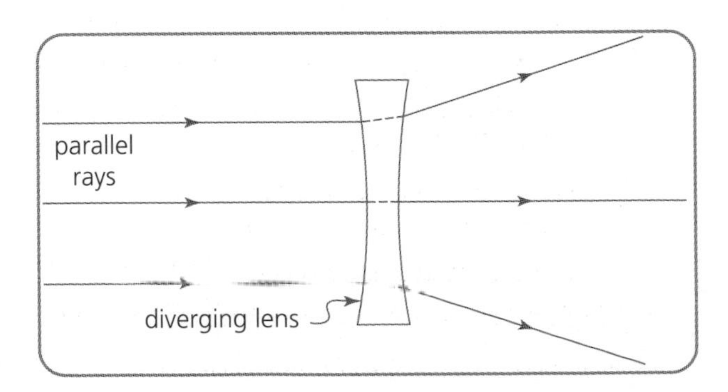

The power of a lens

The power of a lens is a measure of its effect on parallel rays. The greater the power of a lens, the greater the effect it has on parallel rays.

The power of a lens is calculated from

$$P = \frac{1}{f} \quad \text{where:}$$

P is **power** of the lens, measured in *dioptres* (D)

f is the **focal length** of the lens, measured in *metres* (m).

Long and short sight

A long-sighted person sees objects at a distance clearly, but cannot see clearly objects which are only a short distance away.

To see close objects clearly, a *long-sighted* person needs glasses with *converging* lenses.

A short-sighted person sees nearby objects clearly, but cannot see clearly objects which are far away.

To see distant objects clearly, a *short-sighted* person need glasses with *diverging* lenses.

Question and Answer

Explain how a lens is used to correct short sight.

To see distant objects clearly, a short-sighted person needs glasses with diverging lenses.

Rays from objects which are far away are nearly parallel – they diverge very little.

The diverging lens increases the divergence of these rays so that they appear to have come from an object a short distance away.

Exercises

Exercise 20 Refraction

1 Draw a diagram to show the refraction of a ray of blue light incident on the internal surface of a rectangular block of clear plastic.

On your diagram, mark the angle of incidence and angle of refraction.

2 Draw a diagram to show how a converging lens forms the image of an object more than two focal lengths from the lens.

3 Draw a diagram to show how a converging lens forms the image of an object less than one focal length from the lens.

4 A converging lens has a power of 5·0 dioptres. Calculate the focal length of this lens.

5 Explain how a lens is used to correct long sight.

Chapter 5

RADIOACTIVITY

5.1 Ionising radiations

What You Should Know

For Intermediate 2 Physics you need to be able to:

- describe a simple model of the atom
- understand and use the terms 'ionising radiation' and 'ionisation'
- state the nature of alpha particles, beta particles, gamma rays, and their approximate ranges in air
- state that radiation energy may be absorbed by materials
- compare the ionisations produced by alpha particles, beta particles, and gamma rays
- describe how a radiation detector works
- state that radiation can kill or damage living cells
- describe uses of radiation.

The atom

The following diagram shows a model of the atom. You need to be familiar with this model.

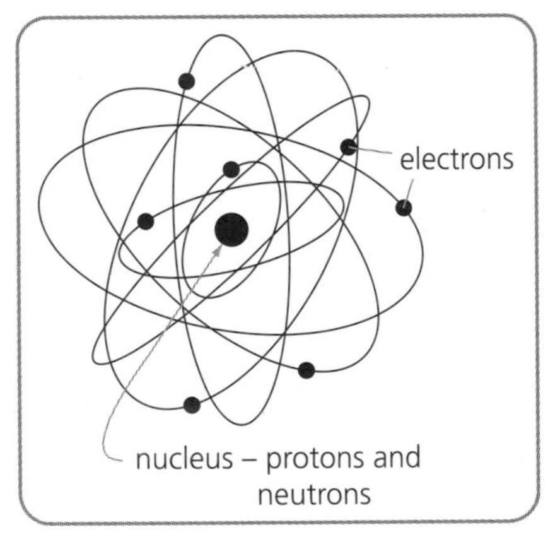

Ionising radiations

For Int2 Physics you need to know about three types of *ionising* radiation: alpha particles, beta particles and gamma rays.

When these radiations pass near atoms they cause ionisation – the atoms gain or lose electrons and become **ions**.

Key Points

1. An *alpha particle* is a cluster of two protons and two neutrons (a helium nucleus). The alpha particle has the largest mass and the biggest charge (+2) and so produces most ionisation. As ions are produced, the energy of the alpha particle is absorbed by the material and the alpha particle is stopped. Alpha particles have a range of only a few centimetres in air.

   The ionisation produced by alpha particles in much greater than the ionisation produced by beta particles or gamma rays.

2. A *beta particle* is a fast moving electron. It has a very small mass and a charge of –1. Beta particles cause more ionisation than gamma rays.

   The energy of the beta particle is absorbed by the material as ions are produced, and the beta particle is stopped. Beta particles have a range of between 20 and 25 centimetres in air.

3. A *gamma ray* is an electromagnetic wave with a very short wavelength and a very high frequency. Gamma rays produce the least ionisation.

   No thickness of any material can guarantee to stop all of the gamma rays from a source.

Detecting radiation

Detectors of radiation operate by using the ionisation which is caused by the radiations.

For example, the film in a *film badge* is blackened by the ionisation caused by radiation. The degree of blackening of the film is a measure of amount of radiation absorbed by the film.

In a *Geiger-Müller tube* (GM tube), ionisation causes a very short pulse of current. Each particle produces a pulse, so that the number of particles going through the tube can be counted.

In a *scintillation counter*, ionisation causes little flashes of light. Each particle causes a flash. The flashes of light are then counted.

Using radiation

Ionising radiations can kill or change the nature of plant, animal and human cells. This effect is used in medicine to sterilise the instruments that surgeons, dentists and others use. The radiation kills any bacteria on the instruments.

Radiation is also used to kill cancer cells in patients.

The ionisations produced by radiations make them relatively easy to detect. This property is often used in medicine and engineering investigations.

For example, a doctor may add a small sample of a radioactive source to a patient's food to investigate how the patient's kidneys are working. The doctor can then test for the presence of the radioactive source in the patient's urine.

Exercises

Exercise 21 Ionising radiations

1 Alpha particles were the first ionising radiation to be discovered. State a reason for this.
2 Draw a diagram to show the structure of an alpha particle.
3 The maximum range in air of beta particles from a radioactive source is 23 cm.
   (a) Explain what happens to the energy of the beta particles.
   (b) Explain whether it would be dangerous to swallow a sample of the source of these beta particles.
4 Doctors are very unlikely to use an alpha source for any investigation inside a patient's body. Why?
5 A water engineer is trying to find underground leaks in the water supply pipes.
   (a) Which type of radiation is most suitable for this type of investigation? Explain your answer.
   (b) Explain how the engineer might use a source of radiation to find leaks.

5.2 Measuring radiation

What You Should Know

For Intermediate 2 Physics you need to be able to:

- define and carry out calculations on the activity of a radioactive source
- define and carry out calculations on absorbed dose, radiation weighting factor, and equivalent dose
- state the factors which affect the risk of biological harm from exposure to radiation
- describe factors affecting background radiation.

Activity of a source

The *activity* of a radioactive source is the number of decays per second, and it is calculated as follows:

$$A = \frac{N}{t}, \quad \text{where:}$$

A is **activity** of the source, measured in *becquerels* (Bq)
N is **number of decays**, just a number and so it has no unit
t is **time**, measured in *seconds* (s).

1 Bq = 1 decay per second.

Absorbed dose

The *absorbed dose* is the energy absorbed per unit mass of the absorbing material, and it is calculated from:

$$D = \frac{E}{m}, \quad \text{where:}$$

D is **absorbed dose**, measured in *grays* (Gy)

E is **energy**, measured in *joules* (J)

m is **mass**, measured in *kilograms* (kg).

1 Gy = 1 J/kg.

Equivalent dose

Each kind of radiation is given a *radiation weighting factor*, w_R, as a measure of its biological effect. The greater the radiation weighting factor, the higher the risk of biological harm.

Equivalent dose is the product *absorbed dose* × *radiation weighting factor*. That is:

$$H = D\,w_R, \quad \text{where:}$$

H is **equivalent dose**, measured in *sieverts* (Sv)

D is **absorbed dose**, measured in *grays* (Gy)

w_R is **radiation weighting factor**, just a number and so it has no unit.

You need to be able to use all three of these relationships to solve problems.

Question and Answer

In an experiment a doctor exposes a piece of tissue of mass 0·2 kg to an alpha source for a period of 30 mins. The activity of the source is 5·0 MBq. During the experiment the absorbed dose received by the tissue is $4{\cdot}0 \times 10^{-4}$ Gy.

Radiation weighting factor for alpha particles = 20

(a) How many decays occur in the alpha source during the experiment?

(b) Calculate the energy absorbed by the tissue sample.

(c) Calculate the equivalent dose received by the tissue sample.

(a) $A = 5{\cdot}0 \text{ MBq} = 5{\cdot}0 \times 10^6 \text{ Bq}$ $\qquad A = \frac{N}{t}$

$t = 30 \text{ mins} = 1800 \text{ s}$ $\qquad \Rightarrow 5{\cdot}0 \times 10^6 = \frac{N}{1800}$

$N = ?$ $\qquad \Rightarrow N = 9{\cdot}0 \times 10^9.$ *(no unit!)*

Question and **Answer** *continued* ➢

Question and Answer continued

(b) $D = 4{\cdot}0 \times 10^{-4}$ Gy

$m = 0{\cdot}2$ kg

$E = ?$

$$D = \frac{E}{m}$$

$$\Rightarrow \quad 4{\cdot}0 \times 10^{-4} = \frac{E}{0{\cdot}2}$$

$$\Rightarrow \quad E = 8{\cdot}0 \times 10^{-5}\ \text{J}.$$

(c) $D = 4{\cdot}0 \times 10^{-4}$ Gy

$w_R = 20$

$H = ?$

$$H = D\,w_R$$

$$\Rightarrow \quad H = 4{\cdot}0 \times 10^{-4} \times 20$$

$$= 8{\cdot}0 \times 10^{-3}\ \text{Sv}.$$

Risk of biological harm

Key Points

The factors which affect the risk of biological harm from exposure to radiation are:

- absorbed dose – the greater the absorbed dose, the higher the risk
- the kind of radiation – the greater the radiation weighting factor, the higher the risk of biological harm
- the body organs or tissue exposed.

You need to know about these factors – so learn them!

Background radiation

It is not possible to eliminate radiation from our lives. We are exposed to low levels of radiation every day. We call this *background radiation*.

Key Points

The level of background radiation depends on:

- radioactive sources inside our own bodies – for example some of the carbon and potassium atoms in our bodies are radioactive!
- radioactive sources in the soil, rocks, air, water, and plants, that surround us
- radon gas – fortunately this gas is rare but it is very radioactive; radon seeps out of rocks and can accumulate inside houses and other buildings
- radioactive contamination – for example the level of background radiation is still very high around the former Chernobyl nuclear power station
- cosmic ray activity – charged particles which come mainly from the sun and sources outside our solar system – we have no control over cosmic ray activity.

You could be asked to describe factors which affect background radiation so make sure you know and understand them.

Exercises

Exercise 22 Measuring radiation

1 Define the becquerel.

2 A student places a Geiger-Müller tube and counter close to a source of gamma rays. The counter registers 3600 counts in a two-minute period. The student intends to use these measurements to calculate the activity of the source.

(a) State two reasons why the activity of the source cannot be calculated from these measurements.

(b) Calculate the activity measured by the student.

3 During a working week a technician in a nuclear power plant is exposed to the following absorbed doses of radiation:

$3{\cdot}4 \times 10^{-5}$ Gy of fast neutrons – radiation weighting factor 3

$6{\cdot}1 \times 10^{-5}$ Gy of beta particles – radiation weighting factor 1

$8{\cdot}2 \times 10^{-4}$ Gy of gamma rays – radiation weighting factor 1.

(a) Calculate the total equivalent dose received by the technician.

(b) For which radiation is the technician's risk of biological harm greatest? Explain your answer.

5.3 Half-life and safety

What You Should Know

For Intermediate 2 Physics you need to be able to:

- understand and use the term 'half-life'
- describe how to measure the half-life of a radioactive source
- carry out calculations on half-life
- describe safety procedures for handling radioactive sources
- describe ways of reducing equivalent dose
- identify the radioactive hazard sign.

Half-life

The activity of any radioactive source decreases with time. The half-life is the time it takes for the activity of a radioactive source to fall to half its original value. After two half-lives the activity is a quarter, after three half-lives it is an eighth … and so on. You need to be able to carry out calculations on half-life so make sure you understand the solution to question 1 in Exercise 23.

For Int2 Physics, you also need to be able to describe how to measure the half-life of a radioactive source. One method is described next.

Question and Answer

Describe an experiment to measure the half-life of a radioactive source.

1. Set up a Geiger-Müller (GM) tube and counter in the area where the experiment is to be carried out. Zero the counter and note the number of counts recorded in a five-minute period. Work out the count rate in counts per minute – this is a measure of the level of background radiation.
2. Place a radioactive source close to the GM tube, as shown.

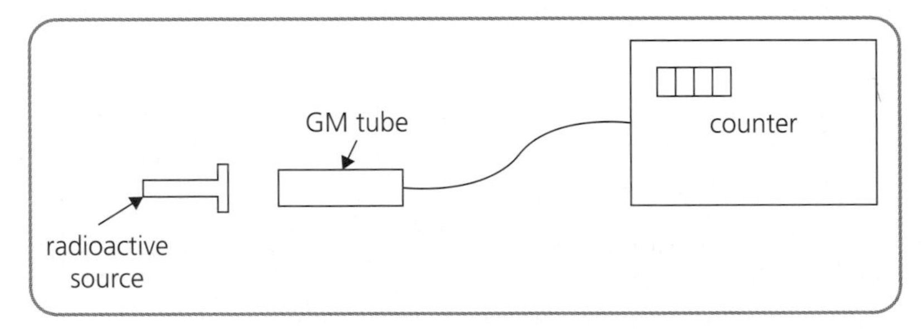

3. Zero the counter and note the number of counts recorded in a one-minute period.
4. Repeat step 3 at five-minute intervals until at least eight measurements have been taken.
5. Subtract the background count rate from each measurement to obtain a corrected count rate for activity due to the radioactive source.
6. Plot a graph of *corrected count rate* vs *time* – the shape of the graph is as shown.

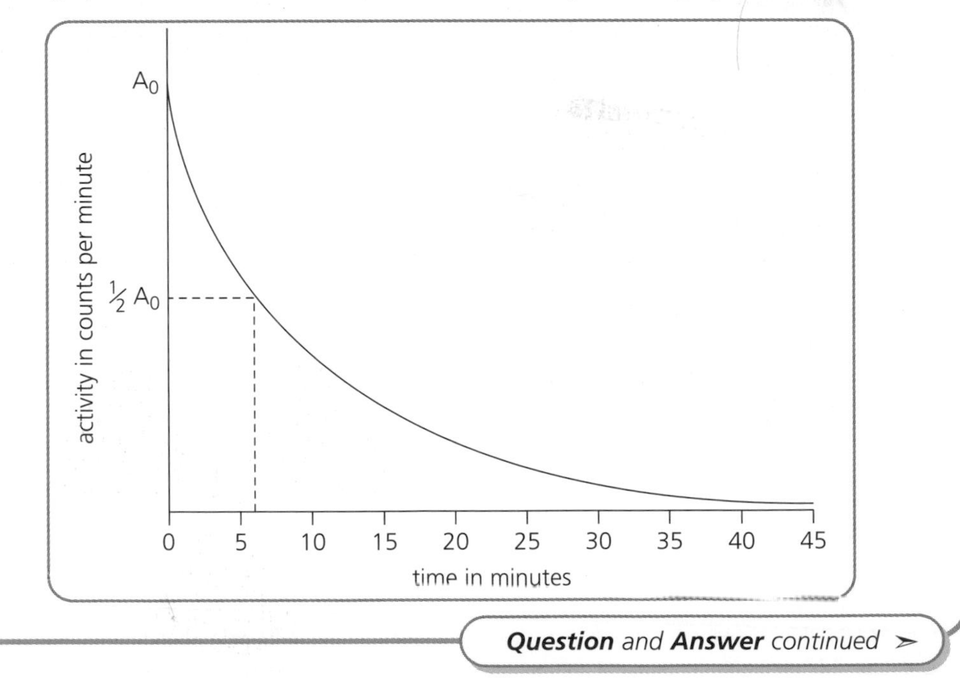

Question and **Answer** *continued* ➢

Question and Answer continued

7 Note the activity, A_0, at $t = 0$ minutes. Calculate $\frac{1}{2}A_0$.

8 Draw a dotted line as shown, from $\frac{1}{2}A_0$ till it meets the graph, then another dotted from this point down to the time axis. Note the difference in time from $A = A_0$ to $A = \frac{1}{2}A_0$.

9 Repeat steps 7 and 8 for the activity values at $t = 1$ minute, 2 minutes, 3 minutes, and 4 minutes.

10 Calculate the average difference in time – this is the measured value of the half-life of the source.

Safety

Take care

For Int2 Physics you need to be aware of precautions that should be taken when handling radioactive sources. For example:

- wearing thick gloves to prevent contamination of your hands
- using tongs to keep your hands at a distance from the source
- wearing protective clothing to prevent contamination of your clothes
- never pointing a source at anyone.

Key Points

You also need to know that *equivalent dose* is reduced by:

- shielding with lead or other materials
- minimising the time of exposure to radiation
- increasing distance from the source.

The radioactive hazard warning sign should be displayed where radioactive sources are used or stored. The sign is as shown – make sure you can identify it.

Exercises

Exercise 23 Half-life and safety

1 In 2008 the activity of a radioactive source is 6·4 MBq. The half life of the source is 40 years. In what year will its activity be $1{\cdot}0 \times 10^5$ Bq?

2 A student is carrying out an experiment to measure the half-life of a radioactive source. State three precautions the student could take.

3 State two ways in which a nuclear scientist can reduce the equivalent dose she receives each week.

5.4 Nuclear reactors

What You Should Know

For Intermediate 2 Physics you need to be able to:

- state advantages and disadvantages of nuclear power
- describe the process of fission
- explain the term 'chain reaction'
- describe how a nuclear reactor works
- describe problems associated with radioactive waste.

Advantages and disadvantages of nuclear power

Nuclear reactors play a major part in the generation of electricity. You should be aware of some of the advantages and disadvantages of nuclear power.

One advantage of using nuclear reactors is that they do not put any greenhouse gases into the atmosphere so do not add to a problem which many scientists believe is causing global warming. Nuclear reactors produce far more energy per kilogram of fuel than power stations fired by fossil fuels. They also let us use less fossil fuels so that more will be left for future generations.

One disadvantage of using nuclear reactors is that they produce radioactive waste products with very long half-lives. These radioactive products will be dangerous for thousands of years or until science finds a way of making them safe.

At the end of its useful life it is very expensive to dismantle (decommission) a nuclear power station.

Some people are very worried about the danger of nuclear accidents.

Storing radioactive waste is a major problem as there are no permanent disposal facilities. When potential sites for radioactive storage are being considered, there is often considerable opposition from local communities which fear the increased risk from radiation, and possible damage to the local economy.

How a nuclear reactor works

You do not need to become an expert on nuclear power stations, but you do need to understand the principles on which they are based.

Fission occurs when a large nucleus splits into two smaller nuclei. A small number of fast neutrons are also released. A fission reaction may be spontaneous or it may be stimulated by the nucleus absorbing a slow neutron, as shown in the diagram.

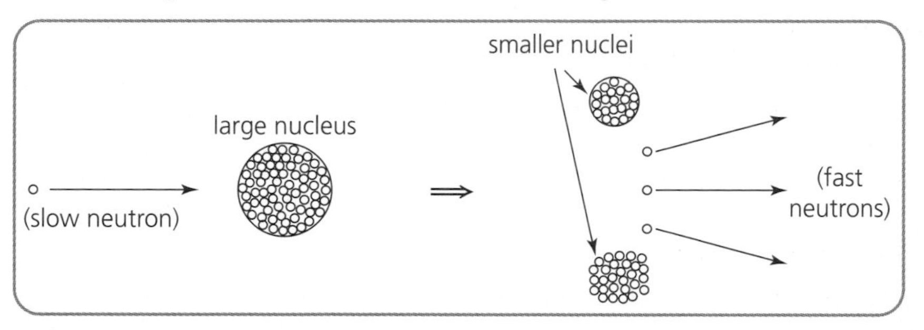

Under the right conditions, neutrons released in one fission may cause further fissions. This is illustrated in the diagram (but without showing the smaller nuclei).

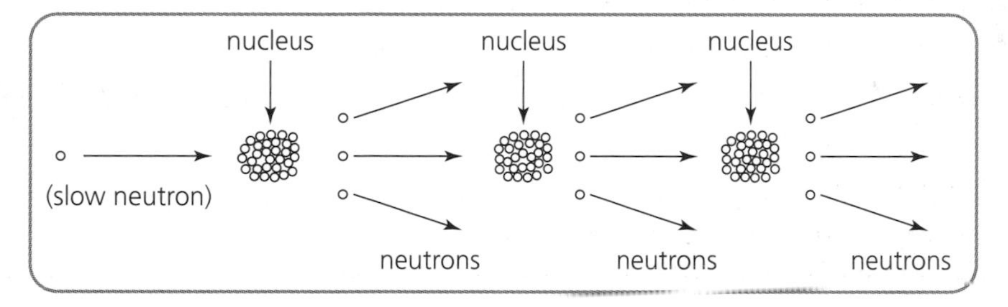

This is called a **chain reaction** – the neutrons act as links between consecutive reactions.

The next diagram is a simplified drawing of the main parts of a nuclear reactor

Key Words and *Definitions*

- In a nuclear reactor the fission reactions take place in the **fuel rods** (made of enriched uranium).
- Fast neutrons which are released are slowed down by the **moderator** (made of carbon).
- **Control rods** (made of boron) absorb extra neutrons to ensure that *on average*, each fission reaction produces one other fission reaction.
- The nuclear reactions release large amounts of heat energy inside the reactor. A **coolant** is pumped through the reactor to absorb the heat energy and to carry it to turbines, which turn the generators, to produce electrical energy.
- All of the parts of the reactor are held within a **containment vessel** which has to withstand the high pressures and high temperatures produced inside the reactor. The containment vessel ensures that none of the radioactive material leaks from the reactor.

Exercises

Exercise 24 Nuclear reactors

1 State one advantage and one disadvantage of nuclear energy.

2 In a nuclear reactor, explain the purpose of:

(a) fuel rods

(b) coolant.

Chapter 6

SOLUTIONS TO EXERCISES

6.1 Mechanics and Heat

Exercise 1 Speed, distance, and time

1 (a) $\bar{v} = ?$ $\qquad d = \bar{v}\,t$

$d = 360$ m $\Rightarrow$ $360 = \bar{v} \times 180$

$t = 180$ s $\Rightarrow$ $\bar{v} = 2{\cdot}0$ m/s.

(b) $\bar{v} = ?$ $\qquad d = \bar{v}\,t$

$d = 360$ m $\Rightarrow$ $360 = \bar{v} \times 120$

$t = 120$ s $\Rightarrow$ $\bar{v} = 3{\cdot}0$ m/s.

2 $d = ?$ $\qquad d = \bar{v}\,t$

$\bar{v} = 25$ m/s $\qquad = 25 \times 1800$

$t = 30$ mins $= 1800$ s $\qquad = 45\,000$ m (= 45 km).

(*You may leave the final answer in metres – you should convert to km only if the question asks for the answer in km.*)

3 $t = ?$ $\qquad d = \bar{v}\,t$

$\bar{v} = 3{\cdot}8$ m/s $\Rightarrow$ $11\,400 = 3{\cdot}8 \times t$

$d = 11{\cdot}4$ km $= 11\,400$ m $\Rightarrow$ $t = 3000$ s (= 50 mins).

(*You may leave the final answer in seconds – convert to minutes only if the question asks for the answer in minutes.*)

4 (The girl wishes her average speed to be 5 m/s for 800 m, so calculate the time to complete 800 m at 5 m/s.)

$t = ?$ $\qquad d = \bar{v}\,t$

$\bar{v} = 5{\cdot}0$ m/s $\Rightarrow$ $800 = 5{\cdot}0 \times t$

$d = 800$ m $\Rightarrow$ $t = 160$ s

The girl has taken 160 s to run the first 400 m. She still has 400 m to run and no time left and so cannot achieve an average speed of 5·0 m/s for the whole distance.

5 (a) The pupils should:

- Use a metre stick to measure the diameter of the ball. At least five measurements should be made and an average calculated.
- Set up a light gate, connected to a computer, level with the centre of the stationary ball. The light gate should be in front of, and close to the ball, and when the boy kicks the ball, he has to make sure that it goes through the light gate.

- Use the light gate and computer measure to the time for the ball to pass through the light gate. At least five measurements should be taken and an average time calculated.
- Find the instantaneous speed of the ball by dividing the average diameter of the ball by the average time taken for the ball to pass through the light gate.

(b) The pupils should:

- Use a metre stick to measure the distance between the ball and target. At least five measurements should be made and an average distance calculated.
- Use a stop watch to measure time.
- Start the watch as the ball is kicked, and stop it when the ball hits the target. At least five measurements should be taken and an average time calculated.
- Find the average speed of the ball by dividing the average distance by the average time.

(c) The instantaneous speed is likely to be greater than the average speed.

Air resistance may cause the ball to slow down after it has been kicked.

6 (a) $v = ?$ $\quad$ $d = vt$

$d = 4{\cdot}0$ m $\quad \Rightarrow \quad 4{\cdot}0 = v \times 0{\cdot}080$

$t = 0{\cdot}080$ s $\quad \Rightarrow \quad v = 50$ m/s.

(b) This is an instantaneous speed – it is calculated over a very short period of time.

Exercise 2 Scalars and vectors

1 (a) Speed and distance are scalars.

(b) Displacement, acceleration and velocity are vectors.

2 Distance has size only; displacement has size and direction.

3 Velocity has size and direction; speed has size only.

4 (a) Force and weight have direction.

(b) Force and weight are vector quantities as they have both size and direction.

5 The boy is not correct. When an object moves in a straight line **without changing direction**, the size of the distance and displacement are always the same. If the object changes direction distance and displacement are different.

Exercise 3 Speed-time graphs and velocity-time graphs

1 (a) The initial speed is not 0 m/s – the graph does not start at $v = 0$.

(b) (i) The speed of the object is increasing between points AB, EF, and FG.

(ii) The speed of the object is decreasing between points CD and GH.

(iii) The speed of the object is constant between points BC and DE.

(c) The final speed of the object is 0 m/s.

2 (a) (i) The velocity is positive between points CH.

(ii) The velocity is negative between points OC and HK.

(b) The velocity of the object is zero at points O, C, H, and K.

(c) The acceleration of the object is:

(i) positive between points BD and JK

(ii) negative between points OA, EF, and GI

(iii) zero between points AB, DE, FG, and IJ.

3

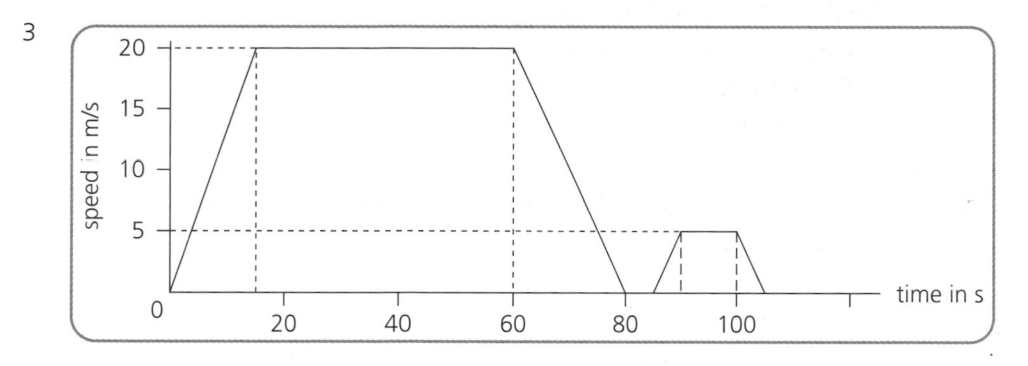

Exercise 4 Graphs, distance, displacement, and acceleration

1 (a) d = area under speed-time graph

= ½ × 5 × 10 (area of triangle)

= 25 m.

(b) d = area under speed-time graph

= 25 + (10 × 10) (add area of rectangle)

= 125 m.

(c) (i) d = area under speed-time graph

= (½ × 5 × 2) + (5 × 8) (area of triangle) + (area of rectangle)

= 5 + 40

= 45 m.

(ii) The cyclist could be slowing down because he is approaching a bend in the track.

2 (a) The ball is dropped at t = 0 s. It falls and bounces at t = 1 s. The ball rises for 0·8 s stops at its highest point momentarily and begins to fall again. The ball is falling at t = 2 s.

(b) Initial height of ball = displacement after 1 s

= area under velocity-time graph

= (½ × 1 × 10)

= 5 m.

(c) Initial acceleration of ball = a during first second

= gradient of velocity-time graph

$= \frac{10 - 0}{1}$

= 10 m/s^2.

(d) Height of first bounce = displacement between t = 1 s and t = 1·8 s

= area under velocity-time graph

= $^1/_2$ × (–8) × 0·8

= –3·2 m.

Thus the height of first bounce = 3·2 m.

(Normally the negative sign must be included in the statement of the final answer; in this case, the wording of the question means that the negative sign may be omitted. If you are in any doubt however include the negative sign.)

(e) Down is the positive direction. The initial movement of the ball is downwards and the initial part of the graph is positive.

3 (a) $a = ?$

$v = 30$ m/s $\qquad a_{(\text{car A})} = \dfrac{v - u}{t}$

$u = 0$ m/s $\qquad = \dfrac{30 - 0}{7{\cdot}9}$

$t = 7{\cdot}9$ s $\qquad = 3{\cdot}79$ m/s^2.

$a = ?$

$v = 25$ m/s $\qquad a_{(\text{car B})} = \dfrac{v - u}{t}$

$u = 0$ m/s $\qquad = \dfrac{25 - 0}{7{\cdot}2}$

$t = 7{\cdot}2$ s $\qquad = 3{\cdot}47$ m/s^2.

Hence the acceleration of car A is greater than the acceleration of car B.

(In this question it is not necessary to round the acceleration values to the correct number of the significant figures (i.e. 2 figures) – these values are not final answers. The final answer is the statement of the greater acceleration.)

(b) (i) $a = ?$

$v = 0$ m/s $\qquad$ *braking* $a_{(\text{car A})} = \dfrac{v - u}{t}$

$u = 24$ m/s $\qquad = \dfrac{0 - 24}{15}$

$t = 15$ s $\qquad = -1{\cdot}6$ m/s^2.

$a = ?$

$v = 0$ m/s $\qquad$ *braking* $a_{(\text{car B})} = \dfrac{v - u}{t}$

$u = 20$ m/s $\qquad = \dfrac{0 - 20}{12}$

$t = 15$ s $\qquad = -1{\cdot}67$ m/s^2.

This suggests that car B has the more effective brakes (larger negative acceleration).

(ii) To make the test fair, both cars should be brought to rest from the same initial speed. The only difference between the tests should be the braking systems of the two cars.

Exercise 5 Force

1 (a) Specimen answers: girl using force of muscles to stretch an elastic band, or fold a piece of paper; weight of a person squashing a cushion when sitting down; wind force bending a tree.

(b) Specimen answers: Force of car engine causing car to get faster; force of gravity making an object fall faster; wind force moving a toy boat faster.

(c) Specimen answers: Footballer using force of (neck) muscles to head a goal; girl using muscle force to hit a hockey ball with a stick; gust of wind catching a falling leaf.

2 Count 50 pencils into a bundle. Tie the bundle together with light thread and attach the bundle of pencils to the hook of a newton balance. Hold the balance vertically and note the reading on the balance. Divide the reading by 50 to find the weight of one pencil.

3 $W = ?$ $\quad W = mg$

$m = 1200$ kg $\quad = 1200 \times 10$

$g = 10$ N/kg $\quad = 12\,000$ N.

4 $m = 1{\cdot}8 \times 10^3$ kg $\quad$ change of weight $= mg_{\text{Earth}} - mg_{\text{Mars}}$

$g_{\text{Earth}} = 10$N/kg $\quad = (1{\cdot}8 \times 10^3 \times 10) - (1{\cdot}8 \times 10^3 \times 4)$

$g_{\text{Mars}} = 4$ N/kg $\quad = 18 \times 10^3 - 7{\cdot}2 \times 10^3$

$= 10{\cdot}8 \times 10^3$

$= 1{\cdot}1 \times 10^4$ N.

5 (a) Air resistance is not helpful when an aircraft is taking off. The force of the air resistance is reduced by making the shape of the aircraft streamlined.

(b) At the start of a race an athlete needs good grip on the ground so that her feet do not slip. The frictional force is increased by making the sole of the running shoe ridged. (For sprinters, sometimes the running shoes have spikes too!)

Exercise 6 Force, resultant force, and motion

1 (a) Balanced – the motion of the pencil is not changing.

(b) Unbalanced – the speed of the pencil is changing as it falls.

(c) Unbalanced – the direction in which the pencil is moving changes.

2 (a) Any stationary object; any object moving at a constant speed in a straight line.

(b) Any object where its speed is changing; any object where its direction of motion is changing.

3 (a) Forward force: engine force. Backward forces: air resistance, force of friction between tyres and the road.

Downward force: weight of car. Upward force: forces between ground and the tyres

(b) Forward force: none. Backward forces: air resistance, force due to brakes.

Downward force: weight of car. Upward force: forces between ground and the tyres.

(c) (i) Horizontal forces are unbalanced. Vertical forces are balanced.

(ii) Horizontal forces are unbalanced. Vertical forces are balanced.

4 (a) *Resultant force*, $F = (8300 - 3200) = 5100$ N.

(b) $m = 3{\cdot}4 \times 10^4$ kg $\quad F = ma$

$F = 5100$ N $\quad \Rightarrow \quad 5100 = 3{\cdot}4 \times 10^4 \times a$

$a = ?$ $\quad \Rightarrow \quad a = 0{\cdot}15$ m/s^2.

5 (First find the resultant of the two applied forces.)

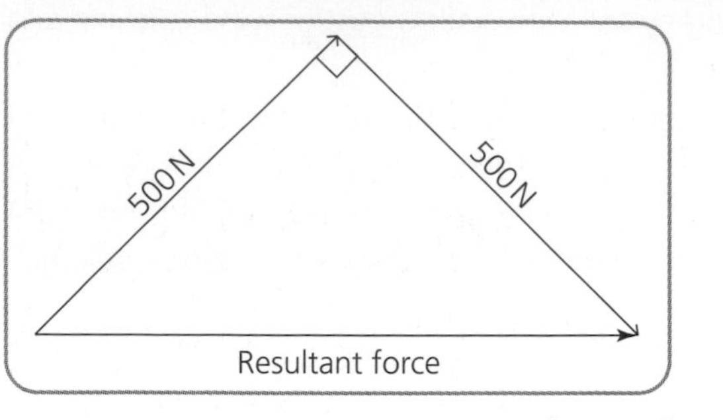

Using Pythagoras Theorem, $F_{(resultant)}{}^2 = 500^2 + 500^2$

$\Rightarrow F_{(resultant)} = 707$

$= 710$ N

Since the crate is moving at a constant speed, the forces acting on it are balanced.

Hence frictional force between the base of the crate and factory floor = (-)710 N.

6 Let the mass of the standard car be m. Then the mass of less powerful car = $0{\cdot}98m$.

Let the force of the standard engine be F. Then the force of less powerful engine = $0{\cdot}95F$.

Acceleration of standard car = $\frac{F}{m}$.

Acceleration of less powerful car = $\frac{0{\cdot}95F}{0{\cdot}98m} = 0{\cdot}969\ \frac{F}{m}$.

Hence the acceleration of the less powerful car is 97% of the acceleration of the more powerful, standard model.

Exercise 7 Projectiles

1 (First calculate the speed of the stone when it reaches the water surface.)

$u = 0$ m/s $\qquad a = \frac{v-u}{t}$

$a = g = 10$ m/s^2 $\qquad \Rightarrow 10 = \frac{v-0}{3{\cdot}0}$

$t = 3{\cdot}0$ s $\qquad \Rightarrow v = 30$ m/s

$v = ?$

(Now sketch the speed-time graph for the stone – a rough sketch is OK.)

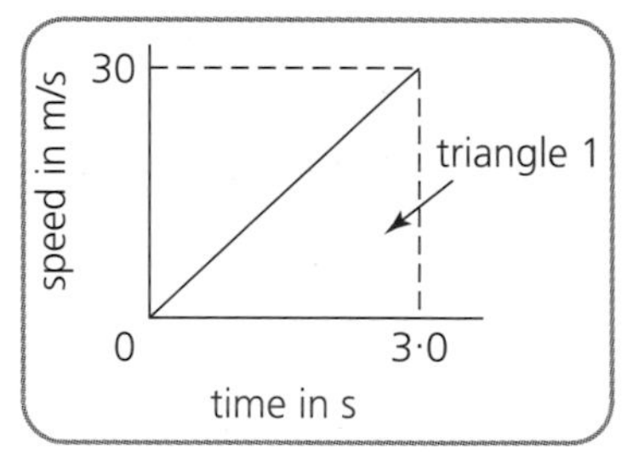

(Now calculate the distance fallen by the stone.)

distance = area of triangle 1

= (½ × 3·0 × 30)

= 45 m

⇒ the water surface is 45 m below the top of the well.

2 (First calculate the time for the rocket to reach its highest point.)

$u = -60$ m/s $\qquad a = \dfrac{v - u}{t}$

$a = 10$ m/s^2 $\qquad \Rightarrow \quad 10 = \dfrac{0 - (-60)}{t}$

$v = 0$ m/s $\qquad \Rightarrow \quad t = 6{\cdot}0$ s.

$t = ?$

Ignoring air resistance, the 'up' and 'down' parts of the rocket's motion take the same time.

Hence the time for rocket to fall from its highest point to Earth = 6·0 s.

(Now sketch velocity-time graph for the falling rocket – again a rough sketch is OK.)

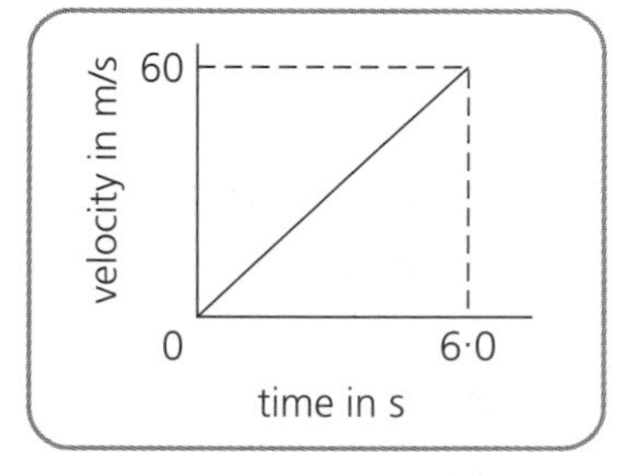

(*The final velocity is +60 m/s i.e. the direction is downwards.*)

Distance fallen by rocket = area under velocity-time graph

= (½ × 6·0 × 60)

= 180 m

Hence the rocket reached a height of 180 m.

3 Ignoring air resistance, the horizontal velocity of the object is constant. In each second the horizontal distance travelled by the object is the same.

The initial vertical velocity of the object is zero and the vertical acceleration is 10 m/s^2. This means that each second the vertical velocity increases by 10 m/s. After 1 s the vertical velocity is 10 m/s; after 2 s it is 20 m/s and so on. In each consecutive second the vertical distance travelled by the object increases.

4 (a) (First use horizontal motion where velocity is constant at 4·8 m/s.)

$t = ?$ $\qquad d = vt$

$v = 4{\cdot}8$ m/s $\qquad \Rightarrow 12 = 4{\cdot}8 \times t$

$d = 12$ m $\qquad \Rightarrow \quad t = 2{\cdot}5$ s.

(b) (Now consider vertical motion.)

$a = g = 10$ m/s^2

$t = 2{\cdot}5$ s $\qquad a = \dfrac{v - u}{t}$

$u = 0$ m/s $\qquad \Rightarrow \; 10 = \dfrac{v - 0}{2{\cdot}5}$

$v = ?$ $\qquad \Rightarrow \quad v = 25$ m/s.

5 It is **not** possible to calculate the horizontal distance travelled by this ball using only time, an initial **vertical** velocity of 0 m/s, and the value of g.

These pieces of information may be used to work out the **initial height** of the tennis ball. The horizontal motion of the ball has no effect on the vertical motion – the time for the ball to reach the ground is the same for any value of horizontal velocity.

Exercise 8 Momentum and Newton's Third Law

1 $p = ?$ $\qquad p = mv$

$m = 1200$ kg $\quad \Rightarrow \quad p = 1200 \times 15$

$v = 15$ m/s $\qquad = 18\,000$ kg m/s.

2

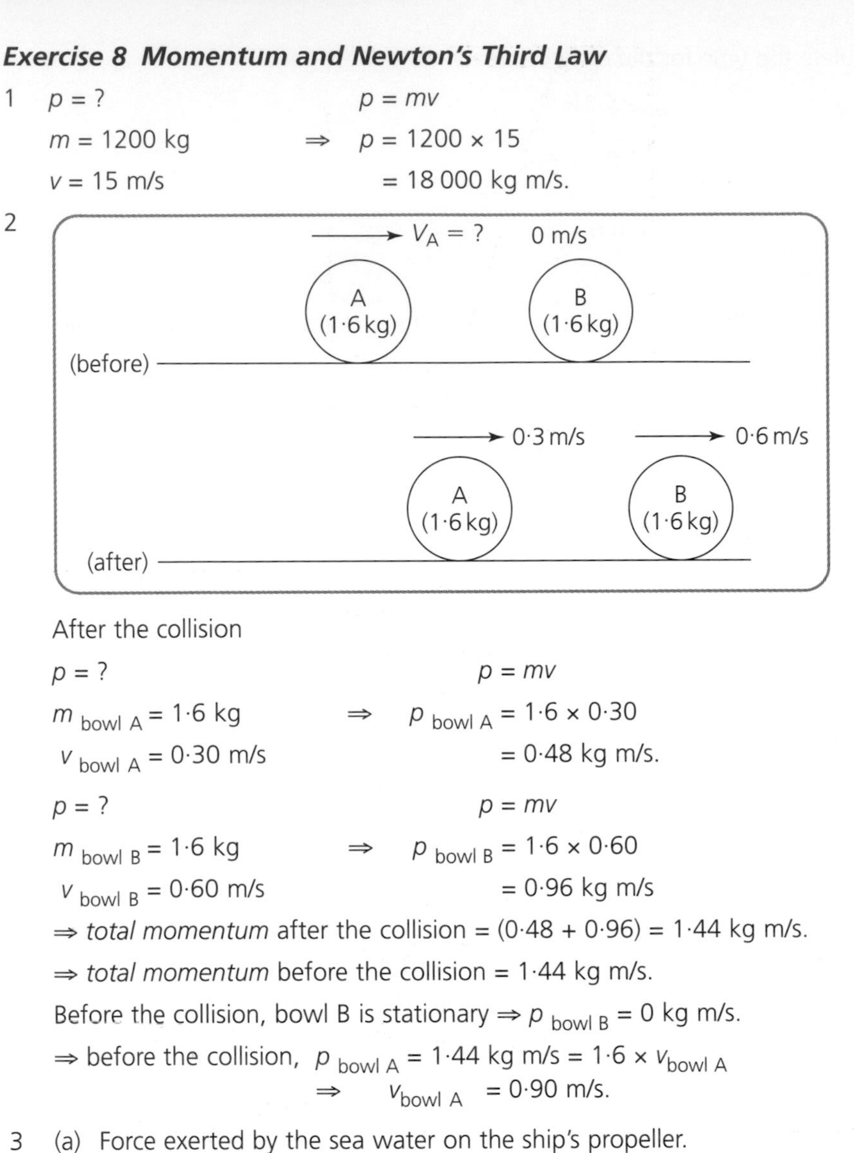

After the collision

$p = ?$ $\qquad p = mv$

$m_{\text{bowl A}} = 1{\cdot}6$ kg $\quad \Rightarrow \quad p_{\text{bowl A}} = 1{\cdot}6 \times 0{\cdot}30$

$v_{\text{bowl A}} = 0{\cdot}30$ m/s $\qquad = 0{\cdot}48$ kg m/s.

$p = ?$ $\qquad p = mv$

$m_{\text{bowl B}} = 1{\cdot}6$ kg $\quad \Rightarrow \quad p_{\text{bowl B}} = 1{\cdot}6 \times 0{\cdot}60$

$v_{\text{bowl B}} = 0{\cdot}60$ m/s $\qquad = 0{\cdot}96$ kg m/s

$\Rightarrow$ *total momentum* after the collision = (0·48 + 0·96) = 1·44 kg m/s.

$\Rightarrow$ *total momentum* before the collision = 1·44 kg m/s.

Before the collision, bowl B is stationary $\Rightarrow p_{\text{bowl B}} = 0$ kg m/s.

$\Rightarrow$ before the collision, $p_{\text{bowl A}} = 1{\cdot}44$ kg m/s $= 1{\cdot}6 \times v_{\text{bowl A}}$

$\Rightarrow \quad v_{\text{bowl A}} = 0{\cdot}90$ m/s.

3 (a) Force exerted by the sea water on the ship's propeller.

(b) The frictional force is a force exerted by the ground on the shoe; the Newton pair force is the force exerted by the shoe on the ground.

(c) The force on the rocket is exerted by the fuel; the Newton pair force is the force exerted by the rocket on the fuel.

4 Forces can only cancel each other out if they act on the same object. Newton pair forces act on different objects.

Exercise 9 Work, energy, and power

1 $d = ?$ $\qquad E_W = Fd$

$E_W = 450$ J $\quad \Rightarrow \quad 450 = 30 \times d$

$F = 30$ N $\quad \Rightarrow \quad d = 15$ m.

2 (The potential energy of the metal sphere changes to kinetic energy as the sphere swings, and the maximum velocity of sphere occurs at lowest point of its swing.)

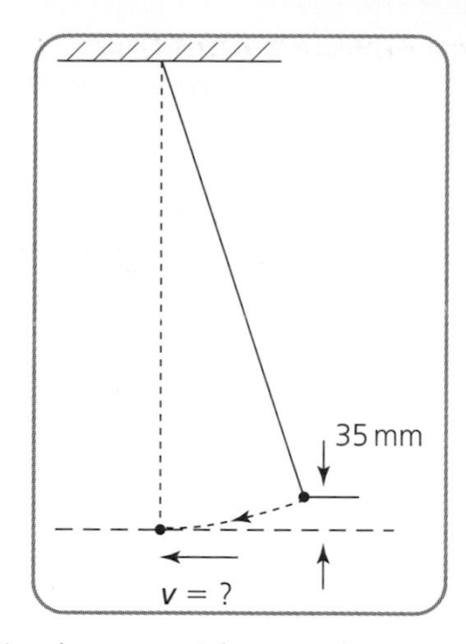

Assume kinetic energy gained = potential energy lost

$\Rightarrow \frac{1}{2}mv^2 = mgh$ (m cancels from both sides of the equation)

h = 35 mm = 0·035 m $\qquad \Rightarrow v^2 = 2 \times 10 \times 0{\cdot}035$

$= 0{\cdot}70$

g = 10 N/kg $\qquad \Rightarrow v = 0{\cdot}836$

$= 0{\cdot}84$ m/s.

3 (a) E = ? $\qquad P = \frac{E}{t}$

P = 12 000 W $\qquad \Rightarrow 12\,000 = \frac{E}{900}$

t = 15 mins = 900 s $\qquad \Rightarrow E = 10\,800\,000$

$= 1{\cdot}1 \times 10^7$ J.

(b) useful E_o = ? $\qquad$ percentage efficiency $= \frac{\text{useful } E_o}{E_i} \times 100$

Efficiency = 30% $\qquad \Rightarrow 30 = \frac{\text{useful } E_o}{10\,800\,000} \times 100$

E_i =10 800 000 J $\qquad \Rightarrow$ useful $E_o = 3\,240\,000$

$= 3{\cdot}2 \times 10^6$ J.

(c) (First calculate the distance travelled by the boat.)

d = ? $\qquad d = vt$

v = 6·0 m/s $\qquad \Rightarrow d = 6{\cdot}0 \times 900$

t = 15 mins = 900 s $\qquad = 5400$ m

Now assume useful 'energy out' of boat engine = work done by engine force.

F = ? $\qquad E_W = Fd$

E_W = 3 240 000 J $\qquad \Rightarrow 3\,240\,000 = F \times 5400$

d = 5400 m $\qquad \Rightarrow F = 600$ N.

4 Consider the relationship for calculating efficiency: *percentage efficiency* = $\frac{\text{useful } E_o}{E_i} \times 100$.

The ratio $\frac{\text{useful } E_o}{E_i}$ is one energy divided by another energy so this part has no unit.

The '100' is just a number and it has no unit ⇒ percentage efficiency has no unit.

Exercise 10 Heat and temperature

1 $E_h = $ | $E_h = ml$
$l = 2{\cdot}25 \times 10^6$ J/kg ⇒ $E_h = 3{\cdot}1 \times 2{\cdot}25 \times 10^6$
$m = 3{\cdot}1$ kg $= 6{\cdot}97 \times 10^6$
$= 7{\cdot}0 \times 10^6$ J.

2 (a) (The calculation is in 3 parts. First the ice heats from –20°C to 0 °C.)

$m = 0{\cdot}72$ kg | $E_h = cm\Delta T$
$\Delta T = 0 - (-20) = 20$ °C | ⇒ $E_h = 2100 \times 0{\cdot}72 \times 20$
$c = 2100$ J/kg °C | $= 30\ 240$ J.
$E_h = ?$

(Secondly, the ice melts.) | $E_h = ml$
$l = 3{\cdot}34 \times 10^5$ J/kg | ⇒ $E_h = 0{\cdot}72 \times 3{\cdot}34 \times 10^5$
$m = 0{\cdot}72$ kg | $= 240\ 480$ J.
$E_h = ?$

(Thirdly, the melt water heats from 0 °C to 15 °C.)

$m = 0{\cdot}72$ kg | $E_h = cm\Delta T$
$\Delta T = 15 - 0 = 15$ °C | ⇒ $E_h = 4180 \times 0{\cdot}72 \times 15$
$c = 4180$ J/kg °C | $= 45\ 144$ J.
$E_h = ?$

Hence the total energy absorbed by ice and its melt water = (30 240 + 240 480 + 45 144)
= 315 800
= $3{\cdot}2 \times 10^5$ J.

(b) The calculation in part (a) assumes that the temperature of the air in the room remains constant.

6.2 Electricity and Electronics

Exercise 11 Circuit basics

1 For charge to flow in a circuit there must be:

- a complete path of conductors
- a source of electrical energy.

2 Conductors are needed to allow charge to flow.

Insulators are necessary so that the circuits can be switched on and off.

3 $I = ?$ $\quad$ $Q = It$

$Q = 600$ C $\quad \Rightarrow \quad 600 = I \times 0{\cdot}15$

$t = 0{\cdot}15$ s $\quad$ $I = 4000$ A.

4

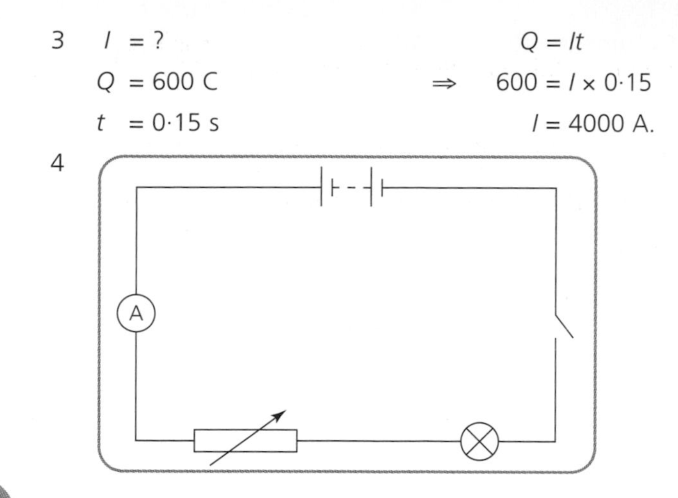

(The ammeter may be connected in series anywhere in the circuit.)

5

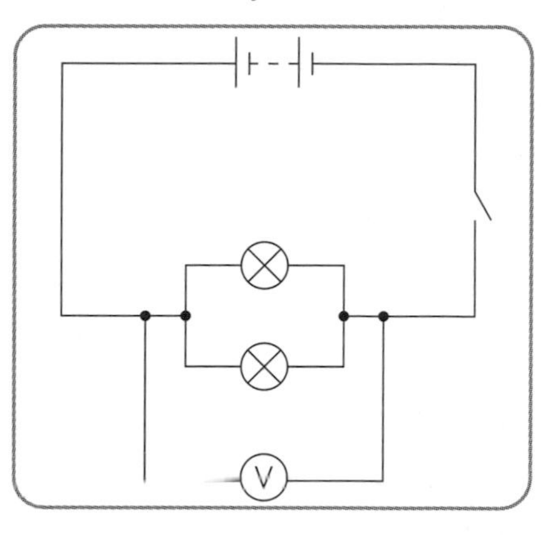

(The voltmeter may be shown anywhere provided that is in parallel with the lamps; the voltage across both lamps is the same.)

Exercise 12 Current, voltage, and resistance

1 (a) (Current in 30 Ω resistor = current in 18 Ω resistor = 0·25 A.)

$V = ?$ $\quad$ $V = IR$

$I = 0{\cdot}25$ A. $\quad \Rightarrow V = 0{\cdot}25 \times 30$

$R = 30\ \Omega$ $\quad = 7{\cdot}5$ V.

(b) R of circuit = ? $\quad$ $V = IR$

$I = 0{\cdot}25$ A. $\quad \Rightarrow 12 = 0{\cdot}25 \times R$

$V = 12$ V $\quad \Rightarrow R$ of circuit = 48 Ω.

(Note: circuit resistance = sum of individual resistances.)

2 (a) $I = ?$ $\quad$ $V = IR$

$V = 6{\cdot}0$ V $\quad \Rightarrow \quad 6{\cdot}0 = I \times 60$

$R = 60\ \Omega$ $\quad$ $I = 0{\cdot}10$ A.

(b) (Since this is a parallel circuit, the current in R_2 = current in battery – current in 60 Ω resistor.)

Hence R_2 current = (0·30 – 0·10)

= 0·20 A.

R_2 = ? $\quad V = IR$

I = 0·20 A. $\quad \Rightarrow \quad 6{\cdot}0 = 0{\cdot}20 \times R$

V = 6·0 V $\quad \Rightarrow \quad R_2 = 30\ \Omega$.

3 (a) The current in the battery = 2·0 A. The battery and the 2·0 Ω resistor are connected in series.

(b) (First calculate the p.d across the 2·0 Ω resistor.)

V = ? $\quad V = IR$

I = 2·0 A $\quad \Rightarrow \quad V = 2{\cdot}0 \times 2{\cdot}0$

R = 2·0 Ω $\quad = 4{\cdot}0$ V.

The path through the 2·0 Ω and the 16 Ω resistors makes a complete path around the circuit.

Hence the p.d. across the 16 Ω resistor = battery voltage – p.d. across 2·0 Ω resistor

= 20 – 4·0

= 16 V.

(c) (First calculate the current in the 16 Ω resistor.)

I = ? $\quad V = IR$

V = 16 V $\quad \Rightarrow \quad 16 = I \times 16$

R = 16 Ω $\quad \Rightarrow \quad I = 1{\cdot}0$ A.

Now current in resistor R_3 = current in 2·0 Ω resistor – current in16 Ω resistor

= 2·0 – 1·0

= 1·0 A.

R_3 = ? $\quad V = IR$

I = 1·0 A. $\quad \Rightarrow \quad 16 = 1{\cdot}0 \times R_3$

V = 16 V $\quad \Rightarrow \quad R_3 = 16\ \Omega$.

Exercise 13 Working with resistors

1 (a) $R_1 = 10\ \Omega$ $\quad R_T = R_1 + R_2 + R_3 \ldots + R_{10}$

$R_2 = 10\ \Omega$ $\quad \Rightarrow \quad R_T = 10 + 10 + 10 \ldots + 10$

$R_3 = 10\ \Omega$ (etc.) $\quad = 100\ \Omega$.

R_T = ?

(b) $R_1 = 10\ \Omega$ $\quad \dfrac{1}{R_T} = \dfrac{1}{R_1} + \dfrac{1}{R_2} + \dfrac{1}{R_3} \ldots \ldots + \dfrac{1}{R_{10}}$

$R_2 = 10\ \Omega$ $\quad \Rightarrow \quad \dfrac{1}{R_T} = \dfrac{1}{10} + \dfrac{1}{10} + \dfrac{1}{10} \ldots + \dfrac{1}{10}$

$R_3 = 10\ \Omega$ (etc.) $\quad = 1{\cdot}0$

R_T = ? $\quad \Rightarrow \quad R_T = 1{\cdot}0\ \Omega$.

2 (a) (First find the resistance of the parallel branches.)

$R_1 = 8{\cdot}0\ \Omega$ $\qquad \dfrac{1}{R_T} = \dfrac{1}{R_1} + \dfrac{1}{R_2}$

$R_2 = 12{\cdot}0\ \Omega$ $\qquad \Rightarrow \dfrac{1}{R_T} = \dfrac{1}{8{\cdot}0} + \dfrac{1}{12{\cdot}0}$

$R_T = ?$ $\qquad = 0{\cdot}208333$

$\Rightarrow R_T = 4{\cdot}8\ \Omega.$ **(Always remember to invert after the addition!)**

(Now add the other resistance.)

$R_1 = 4{\cdot}8\ \Omega$ $\qquad R_T = R_1 + R_2$

$R_2 = 5{\cdot}2\ \Omega$ $\qquad \Rightarrow R_T = 4{\cdot}8 + 5{\cdot}2$

$= 10\ \Omega.$

(b) $I_{\text{battery}} = ?$ $\qquad V = IR$

$V = 12\ \text{V}$ $\qquad \Rightarrow 12 = I \times 10$

$R = 10\ \Omega$ $\qquad \Rightarrow I = 1{\cdot}2\ \text{A}.$

$V_{5{\cdot}2\ \Omega\ \text{resistor}} = ?$ $\qquad V = IR$

$I = 1{\cdot}2\ \text{A}.$ $\qquad \Rightarrow V = 1{\cdot}2 \times 5{\cdot}2$

$R = 5{\cdot}2\ \Omega$ $\qquad = 6{\cdot}24\ \text{V}.$

(Now subtract this from the battery voltage.)

Hence p.d. across $8{\cdot}0\ \Omega$ resistor $= 12 - 6{\cdot}24$

$= 5{\cdot}76$

$= 5{\cdot}8\ \text{V}.$

3 (a) (First find the resistance of the two resistors in series.)

$R_1 = 8{\cdot}0\ \Omega$ $\qquad R_T = R_1 + R_2$

$R_2 = 10{\cdot}0\ \Omega$ $\qquad \Rightarrow R_T = 8{\cdot}0 + 10{\cdot}0$

$R_T = ?$ $\qquad = 18\ \Omega.$

(Now find the resistance of the parallel branches.)

$R_1 = 18\ \Omega$ $\qquad \dfrac{1}{R_T} = \dfrac{1}{R_1} + \dfrac{1}{R_2}$

$R_2 = 9{\cdot}0\ \Omega$ $\qquad \Rightarrow \dfrac{1}{R_T} = \dfrac{1}{18} + \dfrac{1}{9{\cdot}0}$

$R_T = ?$ $\qquad = 0{\cdot}16666$

$\Rightarrow R_T = 6{\cdot}0\ \Omega.$ **(Parallel resistors, so invert after adding!)**

(b) (First calculate the current in the branch containing the $8{\cdot}0\ \Omega$ resistor.)

$I = ?$ $\qquad V = IR$

$V = 24\ \text{V}$ $\qquad \Rightarrow 24 = I \times 18$

$R = 18\ \Omega$ $\qquad I = 1{\cdot}33\ \text{A}.$

(Now calculate the p.d. across the 8·0 Ω resistor.)

$V = ?$ $\qquad V = IR$

$I = 1{\cdot}33$ A. $\quad\Rightarrow\quad V = 1{\cdot}33 \times 8{\cdot}0$

$R = 8{\cdot}0\ \Omega$ $\qquad = 10{\cdot}64$

$\qquad = 11$ V.

4 $R_1 = 80\ \Omega$ $\qquad V_2 = \left(\frac{R_2}{R_1 + R_2}\right)V_s$

$R_2 = 100\ \Omega$ $\quad\Rightarrow\quad V_2 = \left(\frac{100}{80+100}\right)\times 9{\cdot}0$

$V_S = 9{\cdot}0$ V $\qquad = 5{\cdot}0$ V.

$V_2 = ?$

Exercise 14 Electrical energy

1 In an electric fire the energy change takes place in the resistance wire.

2 The energy change in a filament lamp is electrical energy → light + heat.

3 (a) $\qquad P = IV$

(Now $V = IR$. Substitute for V) $\Rightarrow$ $P = I \times IR$

$= I^2R$.

(b) $\qquad P = IV$

($V = IR \Rightarrow I = \frac{V}{R}$. Substitute for I) $\Rightarrow P = \frac{V}{R} \times V$

$= \frac{V^2}{R}$.

4 (a) $P = 3{\cdot}0$ kW $= 3000$ W $\qquad P = IV$

$V = 230$ V $\quad\Rightarrow\quad 3000 = I \times 230$

$I = ?$ $\quad\Rightarrow\quad I = 13{\cdot}04$

$= 13$ A.

(b) $P = 3000$ W $\qquad E = Pt$

$t = 5$ mins $= 300$ s $\quad\Rightarrow\quad E = 3000 \times 300$

$E = ?$ $\qquad = 9{\cdot}0 \times 10^5$ J.

5 (Four identical elements means that the power used by each element $= \frac{1840}{4} = 460$ W.)

$P = 460$ W $\qquad P = \frac{V^2}{R}$

$V = 230$ V $\quad\Rightarrow\quad 460 = \frac{(230)^2}{R}$

$R = ?$ $\quad\Rightarrow\quad R = 115\ \Omega$.

Exercise 15 Electromagnetism

1 Voltage induced in the secondary coil = 0 V. The transformer only operates with a.c.

2 (a) $V_p = 20$ V, $n_p = 100$, $n_s = 400$, $V_s = ?$

$$\frac{V_s}{V_p} = \frac{n_s}{n_p} \Rightarrow \frac{V_s}{20} = \frac{400}{100} \Rightarrow V_s = 80 \text{ V.}$$

(b) $V_s = 80$ V, $I_s = 0{\cdot}40$ A, $R = ?$

$$V = IR \Rightarrow 80 = 0{\cdot}40 \times R \Rightarrow R = 200\ \Omega.$$

3 (Assume the transformer is 100% efficient ⇒
power in primary circuit = power in secondary circuit = 65 W.)

$V_p = 230$ V (mains), $P_p = 65$ W, $I_p = ?$

$$P_p = V_p I_p \Rightarrow 65 = 230 \times I_p \Rightarrow I_p = 0{\cdot}282 = 0{\cdot}28 \text{ A.}$$

4 (a) $I = 2{\cdot}0$ A, $R = 25\ \Omega$, $P = ?$

$$P = I^2R \Rightarrow P = (2{\cdot}0)^2 \times 25 = 100 \text{ W.}$$

(b) $I_s = 2{\cdot}0$ A, $n_p = 960$, $n_s = 480$, $I_p = ?$

$$\frac{I_p}{I_s} = \frac{n_s}{n_p} \Rightarrow \frac{I_p}{2{\cdot}0} = \frac{480}{960} \Rightarrow I_p = 1{\cdot}0 \text{ A.}$$

(c) (Assume the transformer is 100% efficient ⇒ $P_p = 100$ W.)

$I_p = 1{\cdot}0$ A, $P_p = 100$ W, $V_p = ?$

$$P_p = I_pV_p \Rightarrow 100 = 1{\cdot}0 \times V_p \Rightarrow V_p = 100 \text{ V.}$$

Exercise 16 Electronic components

1 (a) Energy change in a motor is electrical energy → kinetic energy.

(b) Energy change in a thermistor is heat → electrical energy.

(c) Energy change in a buzzer is electrical energy → sound.

(d) Energy change in an LDR is light → electrical energy.

2 All input devices change some form of energy to electrical energy.

3 (a) Component X is a thermistor.

(b) $R_1 = 5{\cdot}0\ \text{k}\Omega$. $\quad V_2 = \left(\frac{R_2}{R_1 + R_2}\right)V_s$

$R_2 = 2{\cdot}5\ \text{k}\Omega \quad \Rightarrow \quad V_2 = \left(\frac{2{\cdot}5}{5{\cdot}0 + 2{\cdot}5}\right) \times 6{\cdot}0$

$V_S = 6{\cdot}0\ \text{V} \quad = 2{\cdot}0\ \text{V}.$

$V_2 = ?$

(Both resistance values are quoted in kΩ; it is not necessary to change them to Ω before carrying out this calculation. Can you think why? Look at the term in brackets!)

Exercise 17 Electronic switches and amplifiers

1 (a) Component X is an n-channel enhancement MOSFET.

(b) Consider the potential divider circuit. During the hours of darkness the p.d across the LDR is exactly half of the supply voltage. Hence resistance of R = resistance of LDR = 5 kΩ.

(c) As the intensity of light incident on the LDR increases the resistance of the LDR falls. Hence the p.d. across the LDR falls, and the MOSFET switches off.

2 (a) The frequency of the output signal is 250 Hz. It is equal to the frequency of the input signal.

(b) $P_{gain} = 10 \quad P_{gain} = \frac{P_o}{P_i}$

$P_i = 1{\cdot}2\ \text{W} \quad \Rightarrow \quad 10 = \frac{P_o}{1{\cdot}2}$

$P_o = ? \quad \Rightarrow \quad P_o = 12\ \text{W}.$

6.3 Waves and Optics

Exercise 18 Waves

1 (a) The frequency of a longitudinal wave is the number of waves made in one second.

(b) Sound is a longitudinal wave.

2 $d = 7{\cdot}0\ \text{m} \quad d = vt$

$t = 35\ \text{s} \quad \Rightarrow 7{\cdot}0 = v \times 35$

$v = ? \quad \Rightarrow \quad v = 0{\cdot}20\ \text{m/s}.$

$v = 0{\cdot}20\ \text{m/s} \quad v = f\lambda$

$f = 20/10 = 2{\cdot}0\ \text{Hz} \quad \Rightarrow 0{\cdot}20 = 2{\cdot}0 \times \lambda$

$\lambda = ? \quad \Rightarrow \quad \lambda = 0{\cdot}10\ \text{m}.$

3 $v = 3{\cdot}0 \times 10^8\ \text{m/s} \quad v = f\lambda$

$\lambda = 250\ \text{m} \quad \Rightarrow \quad 3{\cdot}0 \times 10^8 = f \times 250$

$f = ? \quad \Rightarrow \quad f = 1{\cdot}2 \times 10^6\ \text{Hz}.$

4 Any two from: infrared, microwave, TV, radio.

Exercise 19 Reflection

1 A curved reflector can be used for sending radio waves long distances.

The source of the radio waves is placed at the focus of a curved reflector.

Radio waves incident on the reflector are reflected as a parallel beam.

The parallel beam carries the radio waves over long distances with little loss of energy.

2 (a) The effect which occurs at point X is total internal reflection.

(b) Total internal reflection does not occur.

The angle of incidence (30°) on the internal surface of the prism is less than the critical angle of the glass (for red light).

Exercise 20 Refraction

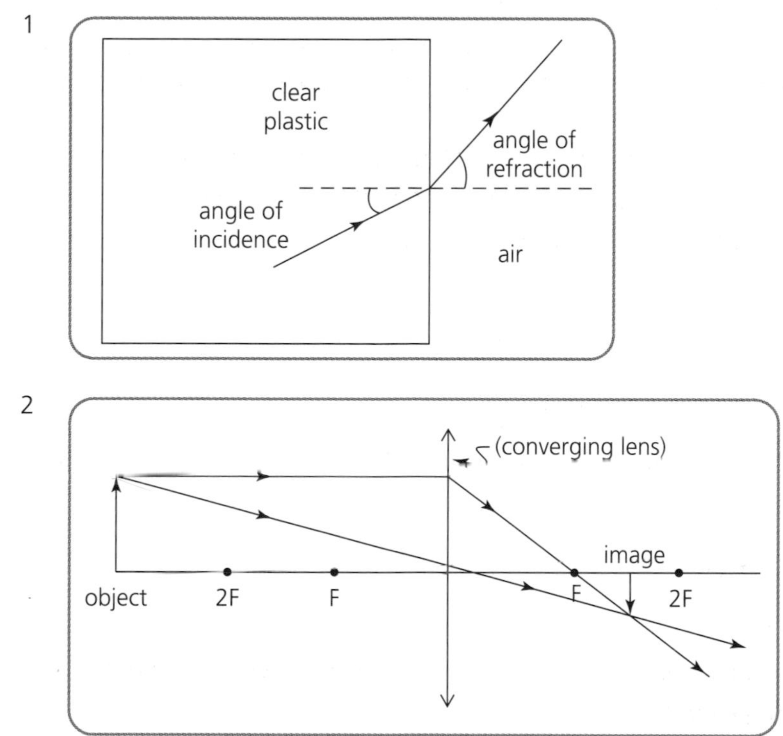

3

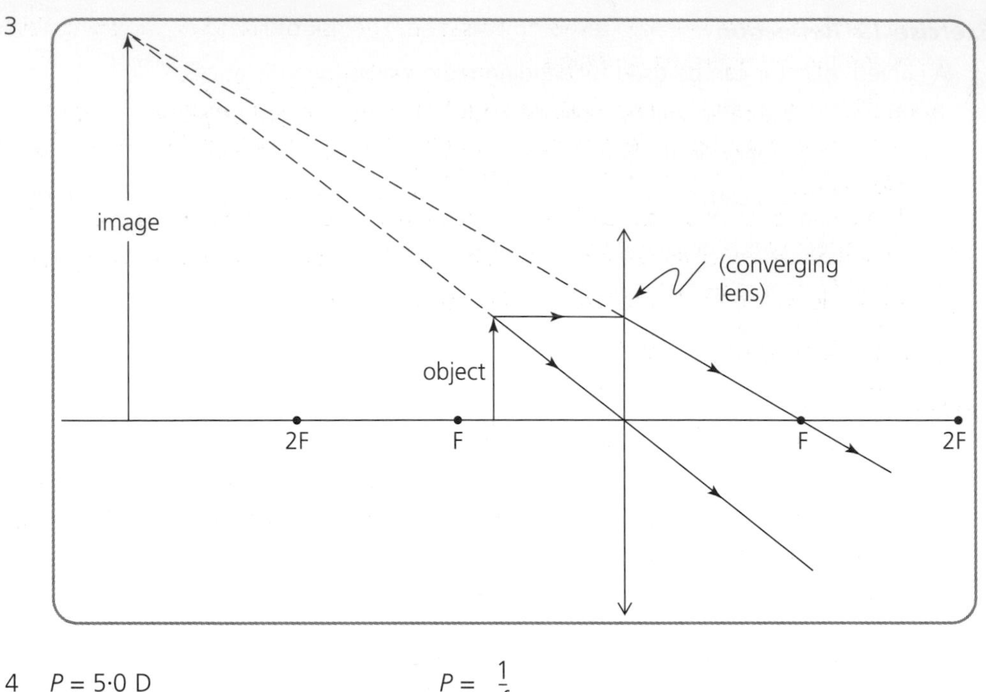

4 $P = 5{\cdot}0\ \text{D}$ $\qquad P = \frac{1}{f}$

$f = ?$ $\qquad \Rightarrow \quad 5{\cdot}0 = \frac{1}{f}$

$\Rightarrow \quad f = 0{\cdot}20\ \text{m}.$

5 A long-sighted person needs glasses with converging lenses to see close objects.

Rays from objects which are close to the eye are diverging too much.

The converging lens reduces the divergence so that that rays appear to come from a more distant object.

6.4 Radioactivity

Exercise 21 Ionising radiations

1 Alpha particles cause far more ionisation than beta particles and gamma rays. This makes them the easiest to detect.

2

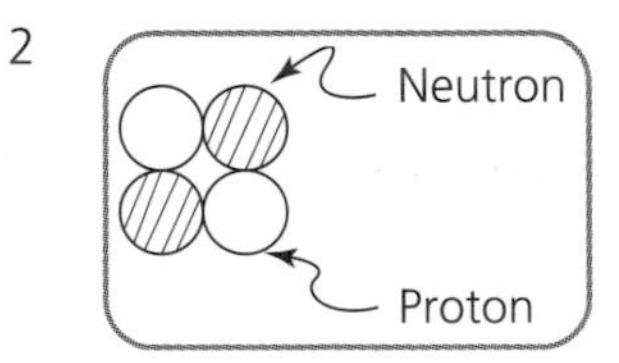

3 (a) The energy of the beta particles is transferred to neutral atoms and causes ionisation.

(b) It would be dangerous to swallow a sample of the source of beta particles.

When the source is inside the body, body tissues in close contact with the source would be exposed to radiation which could kill or harm cells.

4 Alpha particles cause very high levels of ionisation. The risk of killing or damaging tissue inside the patient would be very high.

5 (a) A source of gamma rays is most suitable. Water pipes are underground – alpha particles and beta particles will not reach the surface (they would be stopped by the pipe/earth between the pipe and the surface).

(b) The engineer adds a small sample of a gamma source to the water supply.

S/he then monitors the level of radiation from the source detected at the surface.

At any leak, the level of radiation detected will be higher.

Exercise 22 Measuring radiation

1 In a radioactive source, 1 becquerel is 1 decay per second.

2 (a) (i) The count includes background radiation.

(ii) Radioactive decay is a random process and radiation is emitted in all directions from the radioactive source. The count here includes only particles which pass through the GM tube.

(b) $A = ?$ $\qquad A = \frac{N}{t}$

$t = 2$ mins $= 120$ s $\qquad \Rightarrow A = \frac{3600}{120}$

$N = 3600$ $\qquad = 30$ Bq.

3 (a) (Calculate the equivalent dose for each type of radiation.)

(fast neutrons) $D = 3{\cdot}4 \times 10^{-5}$ Gy $\qquad H = D\,w_R$

$w_R = 3$ $\qquad \Rightarrow H = 3{\cdot}4 \times 10^{-5} \times 3$

$H = ?$ $\qquad = 1{\cdot}02 \times 10^{-4}$ Sv

(beta particles) $D = 6{\cdot}1 \times 10^{-5}$ Gy $\qquad H = D\,w_R$

$w_R = 1$ $\qquad \Rightarrow H = 6{\cdot}1 \times 10^{-5}$ Sv.

$H = ?$

(gamma rays) $D = 8{\cdot}2 \times 10^{-4}$ Gy $\qquad H = D\,w_R$

$w_R = 1$ $\qquad \Rightarrow H = 8{\cdot}2 \times 10^{-4}$ Sv.

$H = ?$

Total equivalent dose $= (1{\cdot}02 \times 10^{-4} + 6{\cdot}1 \times 10^{-5} + 8{\cdot}2 \times 10^{-4})$ Sv

$= 9{\cdot}83 \times 10^{-4}$ Sv

$= 9{\cdot}8 \times 10^{-4}$ Sv.

(b) The gamma rays are the biggest risk – the biggest equivalent dose is for gamma rays.

Exercise 23 Half-life and safety

1 Activity in 2008 = 6·4 MBq = $6{\cdot}4 \times 10^{6}$ Bq

After 1 half life activity will be $\qquad 6{\cdot}4 \times 10^{6} \div 2 = 3{\cdot}2 \times 10^{6}$ Bq

After 2 half-lives the activity will be $\qquad 3{\cdot}2 \times 10^{6} \div 2 = 1{\cdot}6 \times 10^{6}$ Bq

After 3 half-lives the activity will be $\qquad 1{\cdot}6 \times 10^{6} \div 2 = 0{\cdot}8 \times 10^{6} = 8{\cdot}0 \times 10^{5}$ Bq

After 4 half-lives the activity will be $\qquad 8{\cdot}0 \times 10^{5} \div 2 = 4{\cdot}0 \times 10^{5}$ Bq

After 5 half-lives the activity will be $4{\cdot}0 \times 10^5 \div 2 = 2{\cdot}0 \times 10^5$ Bq

After 6 half-lives the activity will be $2{\cdot}0 \times 10^5 \div 2 = 1{\cdot}0 \times 10^5$ Bq.

Now 1 half-life = 40 years, so 6 half-lives = 6 × 40 years = 240 years.

The activity of the source will be $1{\cdot}0 \times 10^5$ Bq in year 2248.

2 Any two from: lead shielding around the source
thick gloves/other protective clothing
using tongs
carrying out the experiment as quickly as safety permits.

3 The nuclear scientist can

(i) use shielding

(ii) minimise the time that she spends in regions with high radiation density.

Exercise 24 Nuclear reactors

1 Advantage: (any one from): no (greenhouse) gases into the atmosphere
more energy per kilogram of fuel
enable conservation of fossil fuels.

Disadvantage: (any one from): radioactive waste products with long half-lives
expensive to decommission
danger of nuclear accidents.

2 (a) The fuel rods contain the material in which the nuclear reactions take place. These reactions release the energy which is used to generate electrical energy.

(b) The coolant absorbs heat energy inside the reactor and carries the energy out of the reactor to the turbines.